PRAIRIE VIEW SUMMER
SCIENCE ACADEMY

AIP CONFERENCE PROCEEDINGS 291

PRAIRIE VIEW SUMMER SCIENCE ACADEMY

PRAIRIE VIEW, TX JUNE 1992

EDITORS:
JOHN F. KRIZMANIC
THE JOHNS HOPKINS UNIVERSITY
JAMES LINDESAY
HOWARD UNIVERSITY

American Institute of Physics   New York

Authorization to photocopy items for internal or personal use, beyond the free copying permitted under the 1978 U.S. Copyright Law (see statement below), is granted by the American Institute of Physics for users registered with the Copyright Clearance Center (CCC) Transactional Reporting Service, provided that the base fee of $2.00 per copy is paid directly to CCC, 27 Congress St., Salem, MA 01970. For those organizations that have been granted a photocopy license by CCC, a separate system of payment has been arranged. The fee code for users of the Transactional Reporting Service is: 0094-243X/87 $2.00.

© 1994 American Institute of Physics.

Individual readers of this volume and nonprofit libraries, acting for them, are permitted to make fair use of the material in it, such as copying an article for use in teaching or research. Permission is granted to quote from this volume in scientific work with the customary acknowledgment of the source. To reprint a figure, table, or other excerpt requires the consent of one of the original authors and notification to AIP. Republication or systematic or multiple reproduction of any material in this volume is permitted only under license from AIP. Address inquiries to Series Editor, AIP Conference Proceedings, AIP Press, American Institute of Physics, 500 Sunnyside Boulevard, Woodbury, NY 11797-2999.

L.C. Catalog Card No. 93-73081
ISBN 1-56396-133-4
DOE CONF-9306394

Printed in the United States of America.

CONTENTS

Preface	vii
Particle Detector Research Center at Prairie View A&M University	ix
Schedule of Speakers	xi
Agenda	xiii
What Good is the SSC? An Introduction to the Physics of Elementary Particles	1
James A. Gates	
Particle Astrophysics or 'Looking' at the Stars	35
Paul Stevenson	
Trends in Data Acquisition Instrumentation	54
George J. Blanar	
Overview of Current High Energy Physics Experiments	68
Larry Gladney	
List of Participants	87

Preface

More than 100 years ago, when the barbaric and distressing idea of slavery was abolished in this country many African-American people now freed had to face such questions as: Who are we? Where are we going? What do we want to be? What should we do for ourselves and our generations to come? In other words, the other face of freedom is creativity. In spite of the advances made through the civil rights movement, an inequality still governs the educational opportunities generally available to minority students. The diminishing number of minorities in the physical sciences indicates that something is dramatically wrong in the social structures which bring them through the K to 12 pipeline.

One purpose of the Summer Science Academy was to address some subset, and not just discuss, this large and growing class of problems affecting the rate and number at which people of color enter the scientific workplace in America.

Contrary to current practice, we chose not to have a summer program for African-American students alone. We wanted to present to them a microcosm mirroring our society, in which future colleagues and possibly future competitors of all races and ethnic origins exist alongside one another.

For the staff here at Prairie View and for professors from the tuples, the Summer Science Academy represented a three day laborfest. For more than 100 visitors coming from all parts of the country, it represented three days of intense lectures on the basics of high energy physics by prominent physicists, theorists, and experimentalists active in the field. It presented an opportunity for professionals from other disciplines to share their thoughts on the importance of technology and its future impact on all of us. For all who made the trip here to Prairie View, the Academy meant a chance to tour research facilities in and around the Houston area. On the final day at Prairie View, it meant a drive north through the beautiful Texas countryside, to a locale just outside of Dallas, a site where we might all meet again, under different circumstances...the Supercollider!

As I watched the students step down from the bus which brought them from Prairie View to the site of the world's largest hadron collider, the SSC, they were indeed a rainbow of colors, countries, and cultures, and the quote engraved on the dome of the Library of Congress came to mind:

> "**One God, One Law**
> **One Element and One**
> **Far-off Divine Event to**
> **Which the Whole of Creation**
> **Moves.**"

Dr. Dennis Judd
Director
Prairie View Summer Academy

Particle Detector Research Center
at
Prairie View A&M University

Directed by the Prairie View A&M University High Energy Physics Group, the Particle Detector Research Center (PDRC) is performing research and development of particle detector technology. This research is needed to analyze the results of the proton collisions in the Superconducting Super Collider (SSC) now under construction in Ellis County, near Waxahachie, Texas. The PDRC is supported by the Texas National Laboratory Research Commission through revenues appropriated by the voters of Texas.

The PDRC Consortium

The PDRC proposal is unique in that it introduces the concept of linking Historically Black Colleges and Universities (HBCUs) with traditionally strong research institutions. These collaborations are building the research infrastructure in the HBCUs while increasing minority contributions to scientific research. Our proposal describes this linkage among consortium universities by comparing it with the mathematical suffix–tuple–which expresses a relationship between two or more elements: "In this Proposal, tuple will take on the meaning of a partnership where the ongoing strength and expertise of one institution is passed to a developing partner in order to achieve a larger goal. Applying this partnership to high energy physics, the name HEP-tuple emerges."

University "Tuples" and Investigators

Prairie View A&M University
(Dennis Judd)

Rice University
(Billy Bonner)

Lincoln University
(Lynn Roberts)

University of Pennsylvania
(N. Lockyer and L. Gladney)

Texas A&I University
(Lionel Hewett)

Texas A&M University
(Russ Huson)

Howard University
(Robert Catchings)

Johns Hopkins University
(Aihud Pevsner)

Southern University
(Chia Yang)

Louisiana State University
(Richard Imlay)

Langston University
(Deborah Gunter)

University of Oklahoma
(George Kalbfleisch)

Superconducting Super Collider (SSC)

The SSC will accelerate protons close to the speed of light in a circular tunnel 53 miles in circumference. The tunnel will contain large superconducting magnets which will collide these subatomic particles creating temperatures and conditions that existed at the beginning of the universe. When completed, the SSC will be the world's largest scientific instrument and will help explain the origin of matter and the basic laws of the universe.

Minority Involvement

The lead institution for the PDRC proposal, Prairie View A&M University, is home to the only high energy physics research group on a predominantly black university campus. The group is headed by Dr. Dennis J. Judd who is one of only a few black high energy physicists in the United States. Dr. Judd states: "Similar programs seeking to increase the numbers of minorities in science and engineering have tended to bring prepared minority students in major universities, but very little money has gone to the minority institutions to build a research infrastructure, which includes trained faculty, better resources, and facilities."

"In this program, minority students won't have to visit other campuses for role models or equipment. They will have access on their own campuses. This is not just another minority 'participation' program. We plan to contribute. We're prepared and we can do it."

Russ Huson (Texas A&M University): "The proposal addresses one of the major problems plaguing our country—that is, how to get minorities involved in science and engineering. This program is different in that it takes a lead, offering solid solutions for improving the physics programs in six minority institutions around the country. This is an enormous improvement over what has been done the past ten years."

PRAIRIE VIEW SUMMER SCIENCE ACADEMY

SCHEDULE OF SPEAKERS

Reception

Tuesday, June 9, 1992

    8:00 p.m. Dennis Judd, Director
Particle Detector Research Center
  at Prairie View
Welcome and Opening Remarks

    8:15 p.m. **Keynote Speaker:**
Martha Smiley, Vice Chairman
Texas National Research Laboratory Commission

Wednesday, June 10, 1992

    9:30 a.m. James Gates, Howard University
"Introduction to High Energy Physics"

    11:00 a.m. Jay Rice, Exxon Corporation
"Careers in Physics Research"

Thursday, June 11, 1992

    9:00 a.m. Paul Stevenson, Rice University
"Particle Astrophysics"

    11:00 a.m. George Blanar, Jr., LeCroy Corporation
"Technology of High Energy Physics"

    7:30 p.m. NCSA SuperQuest Presentation

Friday, June 12, 1992

    9:00 a.m. Larry Gladney, University of Pennsylvania
"Overview of Current High Energy Physics Experiments"

    11:00 a.m. Kate Morgan, SSC Laboratory
"Discover the SSC"

SUPERCONDUCTING SUPER COLLIDER LABORATORY

Prairie View A&M Summer Science Academy

June 13, 1992

AGENDA

8:00 a.m.–9:00 a.m.	Continental Breakfast and Speaker Peter Dingus, Physics Research
9:00 a.m.–9:30 a.m.	Travel to Waxahachie—MDL
9:30 a.m.–11:35 a.m.	Tour of the SSC
11:35 a.m.–12:00 p.m.	Travel to Building No. 4
12:00 p.m.	Lunch with Physicists and Engineers

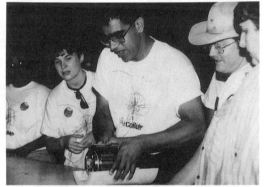

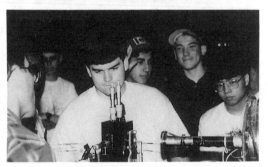

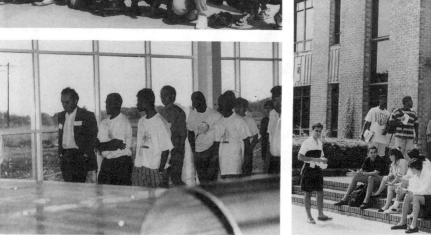

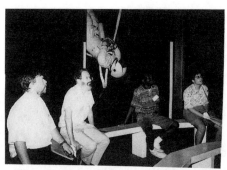

WHAT GOOD IS THE SSC?
AN INTRODUCTION TO THE PHYSICS
OF ELEMENTARY PARTICLES

James A. Gates, Ph.D.
Chairman, Department of Physics
Howard University
Washington D. C.

A FOREWORD AND INTRODUCTION

The universe is a place of wonderful diversity and mystery. Although this sense of wonder is often regarded as the realm of poets, mystics and musicians, it is the task of the scientist to unravel the subtle sign posts that Nature leaves for those of our species with the patience to read them. Through our eyes we see that there are multitudinous complex structures in the world about us. Nevertheless, it has been an underlying theme for first the natural philosopher, later the chemist and more recently the physicist that this apparent fact is just a facade. For over two thousand years the notion that there must exist indivisible bits of matter has been a powerful conceptual theme driving the progress in understanding Nature. At this stage of our scientific development, we now see this theme played out in the context of the physics of "elementary particles."

What is an elementary particle? Well, the simplest way to think about an elementary particle is to imagine a dot. To be a bit more precise, imagine a dot but with no extent. In other words, the elementary particle is best regarded as being the same as the notion of the mathematical point that we learn as the basis of the real number line in arithmetic. However, there is an important difference. Elementary particles possess physically measurable properties such as mass, spin and charge. There is, of course, the additional complication of the fundamental principle of quantum mechanics that tells us that we must simultaneously hold the notion that a point particle (dot) is also a wave-like entity (a wavicle?). I will assume in the following that my reader has made his or her peace with this apparent common sense contradiction.

Eugene Wigner has given an especially appropriate and subtle definition of an elementary particle. Paraphrasing him, we can say that an elementary is an irreducible representation of Lie groups. The important distinction is that he did not say that an elementary particle provides an irreducible representation of group. This raises the question, "What do Lie groups have to do with it?" Or more simply, "What is a Lie group?" In the following we will delve a bit into this question, although we urge the truly interested reader to pursue an independent course of study on this topic.

In the public discourse of this nation, there has been much discussion as to the question from which I draw part of the title of this presentation. In the following I will attempt to give the beginning student a way to start to answer the question. The following is meant to be a gently and largely undemanding introduction to the topic of the physics of the elementary particles.

Before I begin, I wish to thank the organizers of this first meeting of the Prairie View Summer Science Academy at the Particle Detector Research Center here at Prairie View A&M University. My special thanks goes to Professors Dennis Judd and Nigel S. Lockyear for extending the invitation to be the first speaker in what I hope will become a tradition of long standing, excitement and productivity. This is a great honor, although I apologize in advance for perhaps not being able to bring the same excitement that Leon Lederman or Steven Weinberg might have done had either been able to accept their respective invitation to this event. This is the second time to my knowledge that I have been asked to stand in for a well known physicist. The first time was at a meeting of the Gordon Conference in place of Ed Witten. I had a lot of fun on that occasion and I trust that this will be as much fun for me...and my audience as well as the reader.

(I.) THE CAST

A first order of business in gaining an understanding of particle physics is to reach a comprehension of what are the players. Less colloquially, what are the basic constituents of matter and energy? At the present time, all testable evidence suggests that behind the complications apparent to our senses, we can begin a journey down in size to more and more fundamental quantities that are increasing smaller in size.

We begin this journey by recalling the existence of molecules. These are collections of individual atoms and at a typical size of approximately 10^{-8} meters in size. We can imagine going inside the atom to its nucleus. At this point, we are considering structures that are approximately 10^{-15} meters in size. Inside the nucleus there are protons and neutrons. I ask the reader to recall the simple lessons of chemistry, one proton resides in the nucleus of a hydrogen atom, two protons and two neutrons reside in the nucleus of a helium atom, and so forth until we exhaust the entire table of elements by considering ever larger and larger collections of protons and neutrons in the nuclei of heavier and heavier elements.

The other parts of atoms are, of course, the electrons that whirl around the nucleus. For a neutral atom the number of electrons is always equal to the number of protons. Sometimes in very violent collisions of atoms, an electron is 'knocked' completely out of its orbit. When this happens, we say that the atom has been 'singly ionized'. If two electrons are 'knocked' out completely out of their orbits, then the atom becomes 'doubly ionized.' This process can be continued for as long as there are electrons that remain in orbit. In fact, one of the recent developments in physics is the creation of new devices called 'electron beam ion trap machines' or 'EBIT machines.' These devices are incredibly efficient at stripping off electrons but doing this is such a way that the resulting atom is essentially at rest. In the near future it seems likely that such a device will be able to remove all except one in electron for a uranium atom!

However, such familiar objects (protons, electrons and neutrons) are not all 'elementary particles' This raises the question of exactly what is meant by an elementary particle?' The simplest definition that I can give is that an elementary particle is one that cannot be broken up into sub-pieces. This may not seem like a very satisfying definition. But it will suffice as an operational one. With this definition some objects that we think of as being elementary at one time in the history of our species may not be viewed as being elementary at other times. For example, before the discovery of the nucleus, atoms were thought to have been elementary.

At this time, all available scientific data points to electrons as being truly elementary! No experiment to date has yielded any evidence that it is possible to take an electron apart into smaller more elementary pieces. Stated another way there seems to be nothing inside of an electron. One reason for this is that there appears to be no 'inside' for an electron. As far as we can presently tell, it is an example of a 'point particle.' Surprisingly, Nature seems to have played a puzzling trick on us. There is another entity in Nature that looks exactly like an electron except that it is 200 times as heavy. This other quantity is known as an muon. And this puzzle doesn't stop here. There is yet one other particle that also looks exactly like an electron with the difference that is 1700 times as heavy. This last particle is called a "taon" or "tau-particle." As yet a final similar surprise, Nature has provided us with three additional particles ("neutrinos") that are similar to electrons but they have no mass and no charge at all! Collectively all of these six particles are called "Leptons." All seem elementary.

There is an important property that I should make sure that we note. All of the leptons have a property that makes them behave like spinning tops. This behavior is, in fact, called 'spin'. This is a very important property of electrons. Most of the chemical properties of atoms are related to the way which electrons arrange themselves around the nucleus in increasingly large numbers in more and more complicated atoms. To a large degree, the patterns of these arrangements are determined by the spin of the electrons. The spinning of an electron has a peculiar feature that distinguish it from ordinary objects. For example, an ordinary spinning object can be made to spin at different speeds. We see this all the time as we watch the tires on a car. When the car is at rest, the tires are not spinning. As it speeds up, the tires begin to spin faster and faster. An electron, on the other hand, can only speed at two rates! This rate is characterized by a special number that physicists call $\hbar$. An electron can only spin in one direction at a rate of one-half $\hbar$ or in the opposite direction at a rate of one-half $\hbar$. No other rates of spin have ever been observed for electrons. Using the analogy with car tires, if the tires of a car were electrons, then that car could only go forward at one speed or backward at that same speed. But it could never be at rest!

All leptons, have the same rate of spin of one-half $\hbar$. Any entity that has such a rate of spin is called a "fermion." Fermions have one other interesting property. In our everyday experience, the fact that no two objects can occupy the same place at the same time is so obvious that it hardly is cause for comment. But we now know that this property is true for fermions due to one of the most important discoveries in quantum mechanics known as the "exclusion principle." It turns out that not every elementary particle obeys the exclusion principle. We will soon see that there are other particles in nature that have spin that are an even integer times the spin for the electron. These other objects are called "bosons." Since bosons do not obey the exclusion principle one can literally put as many as one pleases in the same place at the same time! Stated another way, if electrons were bosons they would all fall to the lowest orbits in atoms and we would not be here because there would not be any chemistry! To visualize spin, we may think of a tiny ball spinning on an axis as shown in Diagram #1.

On the other hand, we have solid experimental evidence that if you 'shake' a proton or neutron hard enough, you can 'hear' something rattling inside. The way that physicists say this is, "Deep inelastic scattering experiments have shown the existence of point-like substructures within the nucleon." These point-like quantities had been anticipated even before the experimental evidence was complete. The physicist Murray Gell-Mann, who won the Nobel Prize for this insight and called these quantities "quarks", had speculated that such quantities would be found in experiments. Thus a proton or neutron seems to be a collection of three quarks. However, there seems to be eighteen different quarks and depending on how we collect groups of them together we can form protons or neutrons or a lot of other different particles! These collections can either consist of three quarks (in which case we call the collection "a baryon") or one quark and an anti-quark (which is called "a meson").

Furthermore, we also note that the seemingly science-fiction notion of "anti-matter" is an actual science fact. So there really are anti-quarks as well as anti-leptons. When brought together, as in the science fiction novels, matter and anti-matter really do annihilate each other. We can visual this in Diagram #2. In the first an approach of an electron (represented by a dot) toward an anti-electron (represented by the x) is shown. In the second, these two have annihilated each other to be replaced by a "fireball" of energy. The amount of energy in the fireball can be calculated (we assume the approach was very slow) from the most equation of physics given to us by Albert Einstein, E (energy) equals to the m (mass) of a particle times the square of the speed of light ($E=mc^2$).

So apparently the quarks and leptons are the fundamental quantities from which it is possible to construct an entire universe. When we consider quarks and leptons together, we call them "matter." This is appropriate since all the ordinary matter that we see in the universe about us is composed of these fundamental bits. We summarize these facts in the form of two tables (Diagram #3 and Diagram #4).

There is another portion in the cast of characters that we need to "glue together" our universe. It is a fact that in order to construct complicated structures, the basic matter (quarks and leptons) must be assembled into collections. To do this, it is necessary that some "fundamental forces" act on the matter to cause it to aggregate together. It is now believed that we know what these fundamental forces are. There seem to be four such forces; (1.)the gravitational force, (2.) the weak force, (3.) the electromagnetic force and (4.) strong force (Diagram #5).

A remarkable fact is that these forces seem to be associated with particles that are unlike the leptons and quarks that we have discussed previously. We call these other type of particles "gauge fields" or more simply "force carriers." Gauge fields have a very important role to play.

It is well known that like charges repel. This phenomenon of one object causing some effect on another object far away is called "instantaneous action-at-a-distance." Newton for example used this concept to explain how the force of gravity of one heavenly body can affect the motion of another body. However, the concept of instantaneous action-at-a-distance contradicts another physical principle that has a great amount of experimental support. This other principle is the principle of special relativity given to us by Einstein.

Our best theories describing Nature suggest that there is another way in which we can envision why one electron repels another. This is where the notion of a gauge field comes in. Imagine that when an electron is places at one position in space, it sends out a sort of "message carrier" that tells all other electrons where the first electron is located. And further carries the message that these other electrons should move away from the location of the first electron. This is the role of a gauge field. In fact, we now believe that this is exactly how the electromagnetic force of repulsion between two electrons works. We even have a name for the gauge field that carries the information. It is called a "virtual photon." The real photon may be familiar to the reader. It is the "particle of light" that bounces off objects and travels to our eyes thus permitting us to see the object. A virtual photon has some what different properties so we cannot actually see them. But nevertheless the photon is the object that allows all electric and magnetic effects to be transmitted to charged objects like electrons.

Recall that we earlier noted that there are presently known to be four different and distinct fundamental forces. Applying the lesson that we just learned above, we should expect that there should be 'gauge fields' associated with each one of these forces. This, indeed, turns out to be the case. In fact there are lots and lots of force carriers. Let me begin with the electromagnetic force (one more time). It's force carrier is known as the photon. The photon also has a spin. Its rate of spin is twice that of the electron. From our earlier discussion we know that this implies However, unlike the electron, the photon has no mass! We also have already noted that the photon has a spin-rate that is twice that of the electron so it is a boson. This is something that the photon shares with almost all of the different kinds of gauge fields.

As there are other forces, there are other force carriers. For the weak nuclear force, there are three such objects. These are known collectively as the weak intermediate vector bosons. They are referred to individually as the positive W-particle, the negative W-particle and the Z-particle. These objects are incredibly heavy compared to the electron, over one-hundred and fifty thousand times heavier! The weak nuclear force has affects that we can see outside of the laboratory. For example, from science fiction movies (or other such authoritative sources) most of the readers may know that uranium glows in the dark. The process by which this mainly occurs is known as beta-decay. Beta-decay is a direct consequence of the weak nuclear force.

Yet further there is the strong nuclear force and let me describe how anyone can deduce the existence for such a force. The nucleus of the helium atom contains two positively charged protons within a small distance of each other. Since these have the same charge there must be a repulsion between them. If there were no other forces acting, the two protons would tend to pushed away from each other. Since the helium nucleus does not just fly apart normally, something must be holding the protons in place. This something is the strong nuclear force.

More accurately we think the strong force is the residue of another force...this other force is called the "color force" or chromodynamic force. In fact, we have much scientific evidence that the color force acts directly on the quarks that are inside the protons and neutrons. It is thought that the color force is responsible for keeping the quarks together in the interior of hadrons. It does such a good job of this that is seems as though quarks are actually trapped in these interiors. The color force also has a set of carriers, eight of them in fact. These eight objects are called "gluons." This whimsical name is in recognition of the fact that these gauge fields act as the "glue" to keep the quarks inside. Like the photon, the gluons are thought to be massless. We also note that any objects built from quarks, anti-quarks, gluons or anti-gluons are called hadrons.

The rate of spin of the gluons and the intermediate vector bosons are all twice that of the electron. But there is one more force that we haven't discussed. One way in which this force is different from all the others is that the force carrier for this force has a spin-rate that is four times that of the electron. This last force is by far the most familiar of all. For even without being scientists, it is almost universal knowledge that a force is responsible for keeping us on the surface of the earth. This same force is responsible for keeping the earth in orbit about the Sun and the moon in orbit around the Earth. Of course, we are talking about the force of gravitation. The force carrier for the force of gravity is called the graviton. Like the photon, the graviton is thought to be massless. It is of interest to physicists to observe the graviton directly. This may occur in the not too distant future.

Let us make an analogy. James Maxwell discovered a set of equations that describe the behavior of charged objects. These are called Maxwell's Equations. When these were discovered, it was noticed that they predicted that electric and magnetic fields could behave in ways that are similar to waves on the surface of water. These "electromagnetic waves" were later observed in the laboratory by Hertz, the German physicist. This might not, at first sight, seem very important in our every day lives. However, these waves were radio waves. Light is, in fact, another form of these waves.

Albert Einstein gave us a set of equations that are similar to those of Maxwell. Where Maxwell was concerned with the force of electromagnetism, Einstein was concerned with the force of gravity. Einstein's equations are referred to as the "equations of general relativity." Although these equations are very much different from Maxwell's, like them they predict that there can exist in Nature waves of gravity. We would like to observe this prediction. In fact, there is planned to be a major effort expended during the decade of the nineties to do this. The United States government is planning to fund the building of a new research facility that will be known as the Laser Interferometry Gravitational Observatory (LIGO) for this purpose.

With our discussion of the graviton, we have completed the description of the second part of our cast of characters. We summarize these in Diagram # 6.

In our introduction to quarks and leptons, we saw that the notion of anti-matter is a scientific reality. This should raise the question of whether the gauge field also have anti-gauge fields? The answer to this is not so simple. Some gauge fields have anti-particles and some are their own anti-particles! For example, the photon is its own anti-particle. This is true also of the Z-boson and the graviton. On the other hand, the W-minus is the anti-particle to the W-plus. Similarly, for the gluons there are separate anti-particles called anti-gluons.

Now that we have met the complete cast, let's build a universe!

(II) HOW TO BUILD A UNIVERSE

In everyday life we do not see the cast of characters that we have just met. Instead we see them only vaguely. Part of the reason for this is that quarks are very "shy." So shy that no physicists has ever observed a quark directly in the laboratory. It is believed that the gluons keep the quarks forever "glued" to the insides of hadrons. Even if this were not so, we would still expect it to be difficult to see quarks. It is difficult enough to see atoms and molecules. Only within the last five years have we developed the technology to actually "see" individual atoms with STM (Scanning Tunneling Microscopy) technology.

So where are the quarks? Well, let's look inside a typical Helium atom. A typical such atom consists of two electrons in orbit about nucleus at its center. In the interior of the nucleus there are two protons and two nucleus. Let's now look in the interior of one of the protons. At that point something remarkable is found...this is the realm of the quarks. In the interior of the proton there are three quarks, two up quarks and one down quark. Each one of these quarks is a different color. The net combination of three different colors has no net color! (There is a rough analogy here with light. Any artist can tell you that combining all colors gives white light.) That is one way to understand why it takes three quarks. For the neutron, it works much the same way. Looking inside the neutron, we find three different quarks, but this time there are two down quarks and one up quark. The electrical charges work out in an interesting way too. The charge of the proton of the proton is the net sum of the quarks inside. Each up quark has a charge of two-thirds and each one down quark has a charge of minus one-third. So two up quarks and one down quark yields a total of plus one...the charge associated with the proton. Similarly two down quarks and one up quark yield a total of zero...the charge associated with the neutron. All of this is illustrated in Diagram #7.

Notice that so far we have only described combinations built from up and down quarks. Nothing stops us from considering combinations built from up, down and strange quarks. Then we get objects that are not just protons and neutrons! In fact, we can start to build new kinds of nuclear particles. Some of these are illustrated in Diagram #8. This type of diagram is called a "weight space diagram." Since we are now only considering three of the different flavors, we associate the name "SU(3)" with this particular diagram. There are other such diagrams associated with SU(3) but we will not be concerned with those presently.

In this diagram, we see that there are actually eight plus one (i.e. nine) different nuclear particles that are associated with the proton and neutron. These other particles come about from picking different combinations of the up, down and strange quarks. For example, the "Sigma-plus" consists of the combination up-up-strange quark combination. In other words, if we start with a proton (up-up-down quark combination) and replace the down quark by a strange quark, then we have literally changed the proton into a Sigma-plus particle! By similar replacements, we can change protons into other particles. Now unlike the proton and the neutron, the other particles in this diagram are not stable. They "live" only a very short while before they decay into other particles. The family of eight particles very closely related to the proton is called an "SU(3) multiplet" of particles. Since there are eight in the proton family, we call that multiplet an "octet" of particles. The other particle (the Lambda particle) that is not so closely, but still somewhat, related to the proton is called a "singlet." In the laboratory, in addition to singlets and octets, there have also been observed decuplets families of hadrons.

Actually, the patterns that emerge by drawing weight space diagrams touch upon a very beautiful branch of mathematics called the theory of Lie algebras. It was invented by the mathematician Sophus Lie who reputedly set out to invent a form of mathematics that could not possibly be of use to the physical sciences. It turns out that Lie's theory is an extremely useful form of mathematics in many areas of physics! "Lie groups" are the ideal mathematical language to describe symmetries that occur in Nature. I would urge the more technically minded reader to perhaps initiate some study in this area.

There is another place where quarks seem to be lurking about in experiments that we perform in the laboratory. In nuclear physics reactions, it is known that violent (but not too violent) interactions between protons and neutrons can produce particles that are called pions. If we could look inside a pion, we would find that there is one quark and one anti-quark. A pion is an example of a meson. Like the proton (which a baryon), a pion has no net color. (Actually there are three different pions, known as the pi-plus, pi-0 and pi-minus.) But this happens in a different way. Since this is a pi-plus particle, there is one up quark and one down anti-down quark inside. Since "up" is a different flavor from "down" we can easily compute the electric charge. This is shown in Diagram #9.

Even though the flavors are different, the color are chosen so that the up quark is blue and the anti-quark is "anti-blue" because we associate an anti-color with anti-quarks. A color and its anti-color cancel each other out. So the pi-plus has no net color.

I wish now to return to a closer look at forces and the force carriers. As I noted earlier, when most of us learn our fundamentals about electromagnetism, we are taught that like charges repel and opposite charges attract. There is mentioned that the electrical force between two charges is proportional to the product of the two charges and inversely proportional to the square of the distance between the two charges. The is the venerable Coulomb's Law whose mathematical expression is in the following diagram. Due to later progress in physics, we learned that there is another picture that we can associate with this.

We picture this in Diagram #10. On the left side, we see a line that represents the path of an electron moving along. At a certain point its direction is changed. When this occurs, the electron emits a photon that travels along until it encounters a second electron travelling along a different path and causes its motion to change. This is indicated by the solid line on the right hand side. The changes in the two paths of the respective electric represent the well-known repulsion between them.

The wavy line in the middle of the diagram represents the photon as it moves from the first electron to the second. There is a form of mathematics that can be associated to this diagram and can be used to "derive" the fact that the Coulomb Law describes the electrical force between the two electrons. The first physicist who used and developed this technique was Richard Feynman. In his honor, we call these "Feynman Diagrams." Now unlike the photons that we create when we turn on a light, we cannot actually see the photons that tell one electron to move away from another. So to distinguish them, we call them "virtual photons." Any process that involves photons is an electromagnetic interaction. When the effects of quantum mechanics are included (see discussion in the next chapter), we finally reach the theory of "quantum electrodynamics" or "QED."

QED is one of the true triumphs of modern physics. It is, perhaps, the best tested mathematical description of Nature we have. There is a laboratory test, called the "g-2 experiment" that allows testing the mathematical theory to extreme levels of accuracy. At present, such tests have shown agreement between the mathematical theory of QED and the experiments to over twelve decimal places of significant figures! There is no other theory known to this author that has been so rigorously tested.

It is important to note that when the photon was emitted, some property of the electron changes. This something was the direction of motion of the electron in our example. But this principle is a more general one. In general, when one of the force carriers is emitted by one of the matter fields, some property of the matter field must change. We will see more of this shortly.

Earlier we pointed out that one of the fundamental forces in Nature is the Weak Interaction. We further noted that one of the simplest ways to observe this nuclear force in action was to observe that naturally radioactive materials often glow due to a process known as "beta-decay." With our just developed view of the role of gauge fields, we now will examine it in more detail (Diagram # 11).

In the diagram we use the symbol N to represent the neutron. Looking inside, we see three quarks (2 down and 1 up). As the three quarks move along, one of the down quarks spontaneously changes into an up quark! This leaves one down quark and two up quarks. This combination is now a proton! At the point where the down quark changed into an up quark a property of the matter particle changed. At precisely this point, there was an emission of a gauge particle represented by the wavy line. However, this time instead of a photon being emitted, what appears is a W-minus particle. This W-particle is extremely heavy (look back at our table) and unstable. It quickly decays into lighter particles. When this happens there appears precisely at that point an electron and an anti-neutrino. These are in the lower part of the diagram represented by the line with arrows attached to it. So all together, what occurs is that we started with a neutron but wind up with a proton, electron and a neutrino! Below the diagram this is represented by an "equation" that
says N goes to P + e + anti-neutrino. Interactions that involve either the W-plus, W-minus or the Z particles are all effects of the weak interactions. Alternately, when quantum mechanics is included, then this becomes the theory of "quantum flavor dynamics" or "QFD".

Other forces work the same general way. We can consider an example that comes from considering how the gluons affect quarks. For this purpose, we look inside a proton (Diagram # 12).

Here on the left hand side of the diagram, there are three quarks inside (2 up and 1 down). However, remember that quarks not only come in different flavors like up and down. There is additionally the color property of the quarks. In fact, the three different quarks are also three different colors. This way there is no net overall color of the proton. For the purpose of our discussion, let the bottom most quark be a blue, the next one a yellow, and the uppermost one red. Now we follow them along as the proton moves. At a certain point the red quarks spontaneously changes into a yellow one. This is one of those changes of property that we discussed. So simultaneously with this change, a gluon must be emitted. This gluon moves along and then encounters the yellow quark that absorbs it and then changes into a red quark. So if we look at the right hand side, the bottom most quark is blue, the next one is red and the uppermost one is yellow. Since the final proton has three different colors, it has no net color overall.

When gluons are exchanged between particles, the interaction is due to the "color force." This is also called the "chromodynamic force." This type of force is believed to play a critical and very strange role. That is gluons are believed to be the culprits behind the "shyness" of the quarks noted near the very beginning of our presentation. This force is believed to be so strong that it totally prevents any quark from ever getting out of the interior of baryons or mesons. The quarks are thereby "enslaved" or chained to these interiors. Stated another way, the force law between two quarks caused by gluons is not inversely proportional to the distance. Instead this force law is believed to be proportional to the distance that separates the two quarks. Trying to pull two quarks far apart gets more and more difficult until in the end the energy contained in the gluon field is so great that it begins to be converted into the creation of new quarks that are closer to original quarks! Thus instead of getting two quarks separated by a large distance, we wind up with more baryons or mesons than when we began. So quarks never seem to get out!

However, the gluons seem to have another less sinister role than trapping some poor hapless matter particle. These same gluons seem to be the reason that the strong nuclear force exist. Including effects from quantum mechanics and gluons then leads to "quantum chromodynamics."

The astute reader will note that we have not included the final known fundamental force, gravity, into the picture of how such forces work. There is a very good reason for this. Compared to the other fundamental forces, there is a serious lack of understanding of the force of gravity. Although Albert Einstein gave us a stunningly beautiful theory describing many aspect of the gravitational force, his mathematical description called "general relativity", ultimately it does not seem compatible with quantum mechanical theories. Until recently, there seemed to be no way to resolve this dilemma. This changed in the mid-eighties with the realization of the existence of a new class of theories, known as "superstrings" and "heterotic strings." However, a discussion of these esoteric monsters takes us very far afield from our goals in our presentation. Perhaps at some meeting of this school I will have such an opportunity.

(III.) THE STANDARD MODEL, QUANTUM MECHANICS, AND TOWARD THE SUPERCOLLIDER

In this final section, we will discuss what the supercollider can do for us. This will answer the question posed in the title of this presentation. There are several ways in which scientists plan to use the supercollider.

There is a mathematical framework in which all the matter and gauge particles that were introduced in the last section are given a consistent description. This framework is called "The Standard Model" and it is one of the most successful mathematical models of physical phenomena ever created. This model literally agrees with thousands of experiments in nuclear and particle physics! It is one of the best tested theories of physical reality our species has developed. There are many mathematical ingredients that go into the standard model. These include calculus, vectors, spinors, and matrix algebra. For some (like me) who enjoy such matters, there is much fun involved in this. The standard model has a very large number of interesting mathematical properties. However, since this is meant to be a gentle introduction, we will discuss these in general and allegorical terms.

There are also number of puzzling features in the standard model. Some of these include the fact that in addition to matter and gauge particles, its mathematical consistency demands the existence of a somewhat mysterious particle known as the "Higgs particle." This particle is believed to be responsible for the mass of all the previous particles that we have discussed thus far in this article. The history of the concept of the Higgs particle is interesting. Briefly, if we look back at the force carrier particles, we note that only the W-plus, W-minus and Z-particle are the only gauge particle that have a mass. For a long time, there was no consistent way known to physicists how to mathematically describe this situation. This changed when three physicists, Sheldon Glashow, Abdus Salam and Steven Weinberg, were able to discover a way to give this description. The discovery is known as the class of "spontaneously broken gauge theories", and it presently provides the only proven mathematically consistent description of a massive gauge particle. However, this very mathematical consistency demands that the existence of (at least one) Higgs particle. So this becomes a prediction of the standard model. Well, the truth of the matter is that the Higgs particle has not ever been seen in any laboratory experiment! In closing, it should be noted that the Higgs particle is also important for another reason. Although we did not mention much about this previously, with the exception of the neutrinos, <u>all</u> matter particles have mass. In the standard model, all these other masses seem to have their origins with the

Higgs particle.

In general, an invariance is an action that leaves an experiment unchanged. There are a number of invariances associated with the standard model. To illustrate this concept some more lets look at a few other invariances. It seems obvious that performing an experiment at the Fermi Laboratory (Fermilab) near Chicago, Illinois and later moving it to perform the exact same in exactly the same manner experiment at the Stanford Linear Accelerator (SLAC) in Stanford, California should give the exact same answer. Here the action is moving the of position of the experiment both in space and time. These observations imply two invariances. The fact that the laws of physics do not depend on the location where we look at them is known to physicists as "spatial translation invariance." Similarly, the fact that the laws of physics do not depend on the time when we look at them is known as "temporal translation invariance." We also call an invariance by the term "symmetry." We say that temporal translation is a symmetry of the standard model.

Symmetries also have another consequence when they are what are known as "local symmetries" or "gauge symmetries." In a sense the very existence of the gauge fields or force carriers owes their occurrence to these symmetries! To understand this, we need to discuss the concept of local symmetries. Suppose we perform an experiment at Fermilab and move the experimental equipment two feet and repeat it. Also we can perform the same experiment at SLAC and move the experimental equipment say seven feet. Now clearly if it is the same experiment, it should yield the same answer. But the important point is that we moved the experimental equipment a _different_ amount of space at the two _different_ locations. The action of moving it is called a "translation transformation." Another example of this, would be to rotate the experimental equipment at Fermilab a different amount from that at SLAC. This would be a "rotation transformation." Whenever we change some aspect of the physics of an experiment by a different amount depending on the location of the experiment, we are performing a "local symmetry transformation" or a "gauge transformation." It turns out that in order to have the physics of such changes to yield the same answers, it is absolutely necessary for there to exist a "gauge particle" associated with the change. These gauge particles are precisely the force carriers that we met earlier.

Whenever there are local symmetries, these necessarily imply conservation laws! This mathematical result was first derived by Emy Noether and is known as "Noether's Theorem." The conservation of electrical charge (it cannot be created nor destroyed) is an example of a conservation law. Another example is energy (it cannot be created nor destroyed but only change its form) which has a conservation law that is called "The Principle of Conservation of Energy." We have learned due principally to the work of two physicists, Robert Mills and Chen Ning Yang, that gauge fields or force carriers are required in order to local symmetries to exist.

One other "strange" feature of the standard model is known as "CP violation." To get some insight into what is this problem, we need first to discuss some "funny" properties in our universe. We need to investigate the meaning of "C" standing "charge conjugation" and "P" standing for "parity inversion."

By far, P is much simpler to understand. Suppose that we make a movie of a physics experiment. After the experiment is over, we can watch this movie. However, let's be a little perverse. Instead of looking directly at the movie screen, we face away from the screen and watch the movie by using a hand held mirror and record what we see. If after watching the movie in this way, we cannot detect any fundamental difference from watching the screen directly then we physicists say the experiment is "parity invariant."

Alternately, the mirror-watched version of the experiment is called "the parity transformed version" of the non-mirror-watched version. It is well known that if you look at a right hand in a mirror, it looks to be a left hand. Or if we have a ball spinning in a right handed way, looking at it in the mirror will show a ball spinning in a left handed way. For this reason we also say that parity changes the handedness of an object. At one time it was believed that all of physics was parity invariant with respect to the physics of elementary particles. Another way to say this is that Nature has no preferred handedness for particle physics.

Later it was realized through experiment Nature does have a preferred handedness! Recall in our list of leptons, we earlier mentioned three mysterious "cousins" to the electrons called neutrinos. These cousins have no charge and mass. Furthermore and more astoundingly, these particles only exist as "left-handed" particles! (By handedness, I mean spinning in a left handed sense.) So the fact that neutrinos only come in one handedness "explains" why parity is a "broken" symmetry.

We can next turn to the property known as "charge conjugation." To understand what is the concept of charge conjugation , we return to a consideration the pi-zero. This particle is composed (mostly) of a linear of combination up and anti-up quarks and down and anti-down quarks. The simple meaning of charge conjugation is that we interchange all particles with their anti-particles and vice-versa. So what happens to the pi-zero if we perform such an interchange? A little thought reveals that nothing should happen! After performing the interchange, a pi-zero turns into the pi-zero. This situation should be compared with the situation of the proton. If we replace the quarks inside the proton by anti-quarks, then afterwards we have an anti-proton. This is totally different particle! Clearly, charge conjugation is not a symmetry of our section of the universe. Otherwise, we would observe equal numbers of protons and anti-protons. This is fortunate for us as we know what happens when protons and anti-protons meet.

After the discovery that Nature is not parity invariant, a large number of experiments suggested that a combined operation called "CP" seemed to replace it as an invariance. A CP-transformation means looking at our mirror-viewed version and simultaneously making the interchange of all particles with anti-particles. However, it turns out that Nature does not have this as an invariance either! There is a system of particles known as the "K-long" and "K-short" or as the "K-zero" and "K-zero-bar" that show that CP is not an invariance. The fundamental reason why this is so remains unknown.

Before we turn to quantum mechanics, we can use our movie one more time to discuss one other property of the standard model. The property is known as CTP invariance. Let us run our movie of the particle experiment backward while watching in our mirror and simultaneously having interchanged all the particles with anti-particles. (Running the movie backward is the meaning of a "T" or time-reversal transformation.) The meaning of CTP invariance is that if we perform all of these changes simultaneously, there is no way to distinguish our universe from this strangely watched one!

Quantum mechanics has a notorious reputation. Probabilities replace the certainties of Newtonian Mechanics. Electron clouds of probabilities replace the definite orbits around the nucleus. Some common sense questions usually asked can no longer even be asked. In everyday circumstances the quantum nature of matter is not apparent nor important. However, at the level of atoms and below for quarks, leptons and gauge fields, including quantum mechanics effects is crucial. Instead of pursuing a mathematical formulation, we will attempt to approach this in terms of an accessible but necessarily allegorical discussion. The key point is to use Feynman diagrams.

In Diagram #10, we learned to view the force of electrical repulsion in terms of the exchange of virtual photons between two different electrons. This was pictured in that Feynman diagram. Now let's describe another possibility. Suppose that after the first photon was emitted from the first electron, it travelled a distance and then disappeared (Diagram #13)! Elementary particles can (and do) actually behave in this way all the time. This photon was carrying some amount of energy and momentum. Therefore, this energy and momentum cannot just disappear. To carry it an electron and an anti-electron (also called a positron) appear. These are represented by the circle in the middle. They come into existence precisely at the point where the photon disappears. After moving a bit through space they come back together. However, when a particle and anti-particle meet they annihilate each other and give off energy in the form of a second photon. This occurs of the left side of the diagram where the circle joins the second photon line. This second photon then travels through space until it encounters the second electron where it is absorbed and causes it to change its position.

Recall earlier we found that the force of electrical repulsion between to electrons was associated with a Feynman diagram. Now we have discovered another Feynman diagram that connects the two electrons but that has a very different structure from the first diagram. This is very important! It implies that the force of repulsion is not just the usual Coulomb law! In particular, in order to calculate the force, we now have to add the effect of our new diagram to our old diagram. This is shown in Diagram #14.

It turns out that by using the mathematics of the Feynman diagrams it is possible to calculate a mathematical function associated with the second diagram. This function will not only depend on the distance between the two electrons, but also on their speeds. So when quantum mechanics is taken into account the force of repulsion between two electrons becomes much more complicated.

There are a couple of points to note. When we compare these two diagrams, the "outside part" (the lines on the far left and far right) are the same. The first diagram is obviously simpler than the second one. Because it occurs even without including the effects of quantum mechanics, it is called "the classical diagram." The next diagram is called a "quantum correction" or "loop correction." It is important to note also that loop corrections are always dependent on powers of $\hbar$. For the quantum correction diagram given turns out to be proportional to the second power of $\hbar$.

Additionally it turns out that it is possible to construct more diagrams that have the same outside structure describing two electrons repelling each other. So in fact the example given is not the complete answer. There are an infinite number of other quantum diagrams for this process! So the effect of quantum mechanics is to teach us that in physical processes we must add in all possible more complicated diagrams to the classical picture. It takes more energy to actually measure these higher order corrections.

Finally we come to one noticeable feature of the standard model. Upon a review, it is clear that all the matter particles upon which forces act are fermions while all of the force carrying particles are bosons! Why is this the case? In fact, no one knows. Theoretical physicists are forever asking such questions. To this one they have proposed a fascinating answer, "Maybe what we have seen so far is not the case and fermions and bosons are treated in a much more symmetrical way." In other words, maybe in the future we will discover that there exist boson and fermion matter particles and that there may exist boson and fermion force carrying particles. This is the basic proposal of a notion called "supersymmetry" meaning that Nature may have a hidden underlying symmetry between bosons and fermions.

At our present level of understanding, supersymmetry would imply a doubling of everything we have discussed so far. (It should be strongly kept in mind that what we are discussing is purely speculative!) For example, if supersymmetry turns out to be correct, associated with the photon (a boson) there must be a fermion (a spin 1/2 $\hbar$ particle) called a photino. Just as the photon carries the electromagnetic force, so would a photino. Similarly, as there are electrons (a fermion), if supersymmetry is correct, then in Nature there must occur a boson (a spin zero) called the selectron. Associated with quarks (fermions) there would be bosons squarks. Associated with the graviton would be a fermion (a spin three-halves $\hbar$ particle) called the gravitino. And would play this type of game for every particle that we have met in the standard model. Again, it should be emphasized that we are going beyond the standard model in this discussion of supersymmetry. There is no experimental evidence to date for the existence of supersymmetry.

However, there are theoretical reasons to believe that the principle of supersymmetry may be lurking somewhere in the future of particle physics. These reasons are technical but one of them is very simple to understand. It appears that supersymmetry is the only presently known way in which the quantum corrections by Luse of the Feynman diagram technique can be included into a theory with gravity. It is because of this present inability of the standard model to include quantum mechanics and gravity that we know the standard model must ultimately fail as a fundamental description of nature.

So how would it be possible to see all of these new particles and verify that our universe is really supersymmetrical? The only ways involve performing scattering experiments at the highest energies available. There are two ways the may come about. In the first we would observe these new symmetric forms of matter and forces directly. This is illustrated in Diagram #15.

At the top of this diagram we show two protons being propelled toward one another with a large amount of energy. The quantity v ("nu") indicated is the ratio of the electron's speed over the speed of light. When the protons collide, they can create a fireball. This is indicated in the middle of the diagram. Temperatures within it have not been seen here since the conditions very soon after the Big Bang. As the fireball cools, it may be possible to create supersymmetric particles. If we are lucky, it may be possible to observe these.

The second way it may be possible to discover supersymmetrical particles is to look back at our quantum corrections. We recall how the loop corrections work, then an obvious generalization is to allow superpartners and anti-partners to run around the loop, if they exist. This will lead to a different set of quantum corrections! Each set of particle and anti-particles contribute to the loops in different ways. Since these corrections are proportional to higher powers of $\hbar$, they are very small. It takes higher and higher energies to probe for these smaller and smaller corrections. So even if we do not see the super partners directly, they may be detected through this indirect means.

This brings us full face to what good is the SSC? As the reader should have noted in our discussion there are a number of puzzles with the standard model. The problems of the origins of CP violation, the possible existence and mass of the Higgs particle, and the possibility of realizing supersymmetry in

Nature are all questions that cannot be answered at the energies that exist in present operating laboratories. There are other questions such as the existence and mass of the top quark (which will soon be under study at Fermilab) or whether "grand unification" with its likely prediction of the decay of the proton is a reality which requires that accelerators of higher energy be used to carry out complicated and delicate new experiments in particle physics.

Are these questions important given the needs of the Nation in so many other pressing areas of the social and economical welfare of the country? I can only answer that science has always returned a large benefit to the society that has made the investment. It is quite conceivable that new forms of matter and energy will be found in the Superconducting Supercollider. We are not guaranteed to be the first to find these. It is very likely that the Large Hadron Collider (LHC) at the European Center for Nuclear Research (CERN) in Geneva, Switzerland will also be looking into this vast unexplored continent. If this Nation does not go ahead with the SSC, then we will have relented on the pioneering spirit upon which this Nation demonstrated so aptly from its creation until it made the effort of going to the moon. Our Nation will simply be spiritually and intellectually poorer for not carrying through in this new space race. The new frontier will be there for others to explore.

Spin

Almost all elementary particles behave like spinning tops.

Spin - 1/2

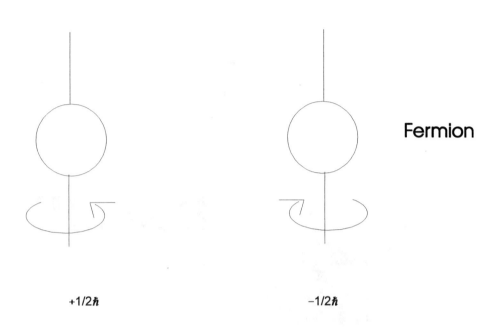

+1/2$\hbar$ −1/2$\hbar$

Diagram 1

But they can only spin at a rate that is an integer or half-integer time $\hbar$

($\hbar$ measure spin angular momentum)

$$\hbar = \frac{h}{2\pi}$$ $h \equiv$ *Planck's Constant*

20 What Good is the SSC?

Anti-particles have the same spin and mass as the corresponding ordinary particle, but all opposite charges. (electric, color, flavor)

But if a particle and anti-particle ever meet.....

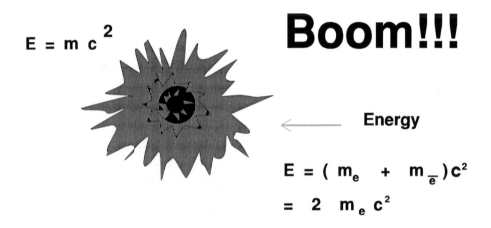

Diagram 2

Basic Cast of Characters

Quarks and Their Charges

Color \ Flavor	u Up	d Down	c Charm	s Strange	t Top	b Bottom
Red	● 2/3	● -1/3	● 2/3	● -1/3	● 2/3	● -1/3
Yellow	● "	● "	● "	● "	● "	● "
Blue	● "	● "	● "	● "	● "	● "

Leptons and Their Charges

Flavor	ν_e Electron neutrino	e electron	ν_μ Muon neutrino	μ Muon	ν_τ Tauon neutrino	τ Tauon
Charge	● 0	● -1	● 0	● -1	● 0	● -1

$$m_\nu \approx 0 (\text{we think!})$$

$$m_e/m_\mu \cong \frac{1}{200} \qquad m_e/m_\tau \cong \frac{1}{1700}$$

Quarks + Leptons are "fermions" with spin = 1/2 $\hbar$

Diagram 3

What Good is the SSC?

Anti-matter is not Science-fiction!

Anti-quarks and Their Charges

Flavor / Color	$\bar{u}$ Up	$\bar{d}$ Down	$\bar{c}$ Charm	$\bar{s}$ Strange	$\bar{t}$ Top	$\bar{b}$ Bottom
Red	X -2/3	X 1/3	X -2/3	X 1/3	X -2/3	X 1/3
Yellow	X	X	X	X	X	X
Blue	X	X	X	X	X	X

Anti-Leptons and Their Charges

Flavor	$\bar{\nu}_e$	$\bar{e}$	$\bar{\nu}_\mu$	$\bar{\mu}$	$\bar{\nu}_\tau$	$\bar{\tau}$
Charge	X 0	X +1	X 0	X +1	X 0	X +1

Diagram 4

Basic Forces (Known)

Weak Interaction: National radioactivity
                                      B - decay

Electromagnetic Interaction: Static electricity, lighting, magnets, electronics

Strong Interaction: Powers the sun
                                 Fusion
                                 Hydrogen bombs

Gravitational Interaction: Keep planets in orbits and you seated in your chair!

Diagram 5

Gauge Fields (force carriers)

Weak Interaction			Mass	Spin
W_μ^+	W - plus		81,000 MeV	
Z^0	Z - boson		91,000 MeV	
W_μ^-	W - minus		81,000 MeV	
$(W^- = \overline{W^+})$				$\begin{bmatrix} \pm 1 \\ 0 \end{bmatrix} \hbar$

Electromagnetic Interaction

       $A_\mu(\gamma)$ photon   0
                                                               $(\pm 1)\hbar$

Strong Interaction

       $G_\mu^j(q)$ gluons   0
                                                               $(\pm 1)\hbar$

Gravitational Interaction

       $g_{\mu\nu}$ graviton   0
                                                                $(\pm 2)\hbar$

Diagram 6

24 What Good is the SSC?

Building up the Nucleon

Proton

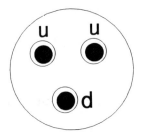

Three quark combination = Baryons

Q_{proton} ΣQ_{quark}

$= e_o (2/3 + 2/3 - 1/3) = +1 \, e_o$

e_o is the basic unit of charge
($Q_{electron} = 1 \, e_o$)

Rule: The sum of three different color is no color at all!

Neutron

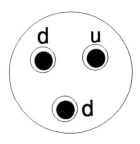

$Q_{neutron} = e_o (2/3 - 1/3 - 1/3) = 0$

Diagram 7

What other combinations are possible?

SU(3)

Up
Down
Strange

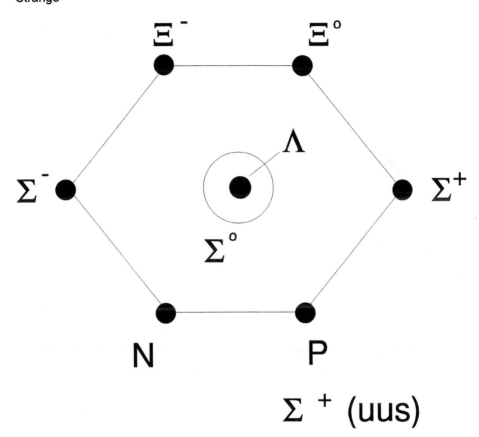

Σ^+ (uus)

Similarly   SU(4) uses   Up
Down
Strange
Charm

Diagram 8

26 What Good is the SSC?

Mesons

quark - anti quark

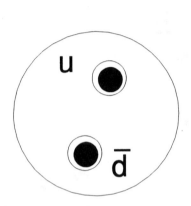
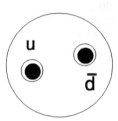
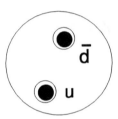

Rule: Anti-quarks have opposite color so quark-anti-quark pair have no color!

$$Q_{pion} = \Sigma\, Q_{quark}$$

$$= e_o[\tfrac{2}{3} - (-\tfrac{1}{3})] = +1\, e_o$$

Diagram 9

Q E D

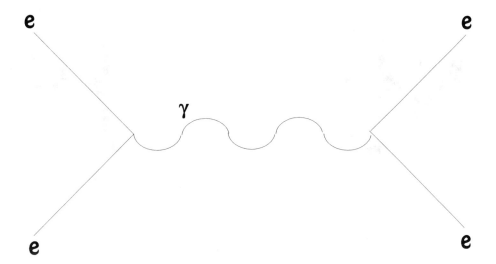

Feynman Diagram

The reason the particle "feel" forces is because the "force carriers" are exchanged between the particles

$$\Rightarrow F = k \frac{e_o e_o}{r^2}$$

Diagram 10

28 What Good is the SSC?

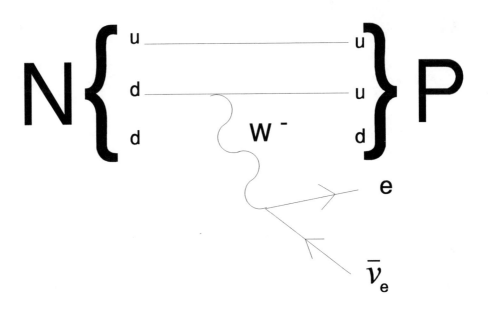

β - decay

N → P + e + $\bar{\nu}_e$

Diagram 11

QCD

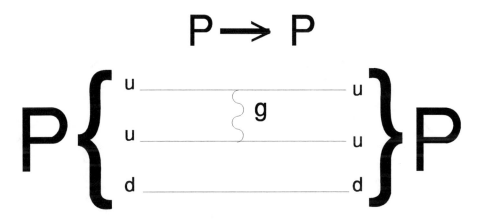

Quarks can change color by exchanging gluons. Gluons keep quarks bound to the interior of baryons and mesons (infra-red slavery, confinement).

Diagram 12

30 What Good is the SSC?

Feynman Diagram View

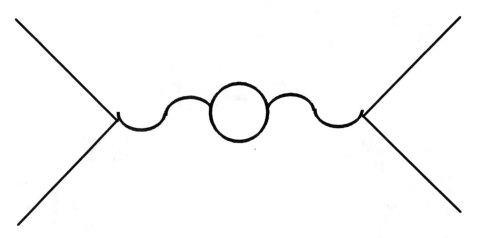

Diagram 13

Checking Q.M.

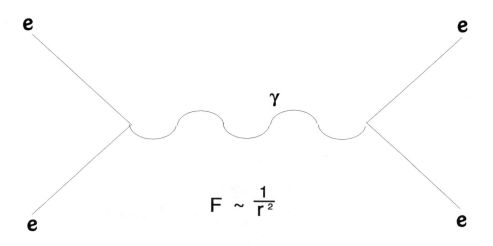

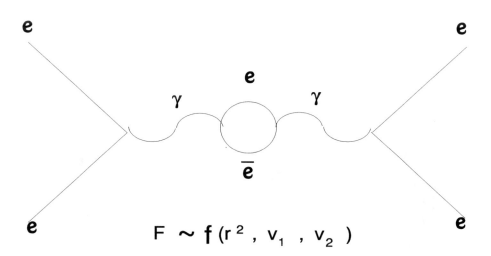

Diagram 14

32   What Good is the SSC?

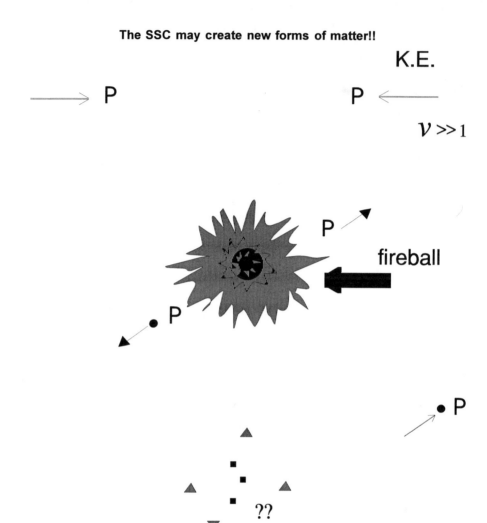

Diagram 15

QCD

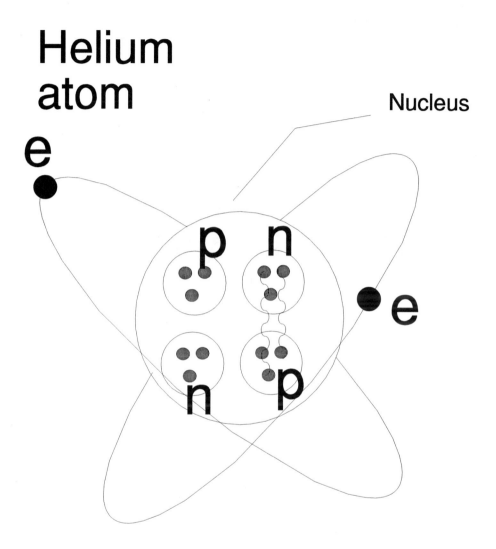

Gluons keep protons and neutrons inside the nucleus (strong interaction).

34 What Good is the SSC?

We are able to write a set of equations that predict in a **consistent** way how the first three forces together with matter behave in our universe. This mathematical construction is called:

"The Standard Model"

$$\mathcal{L}_{sm} = \mathcal{L}_{Dirac} \quad \text{(Matter)}$$

$$\mathcal{L}_{ym} \quad \text{(Gauge fields, Yang-Mills)}$$

$$\mathcal{L}_{Higgs} \quad (??)$$

$$\mathcal{L}_{Dirac} = i\frac{1}{2} Tr[\bar{Q}\nabla Q]$$
$$+ i\frac{1}{2}\Sigma_i \bar{L}_i P_+ \nabla_L L_i$$
$$+ i\frac{1}{2}\Sigma_i \bar{R}_i P_+ \nabla_R R_i$$

$$\mathcal{L}_{ym} = \frac{1}{g^2} Tr[F_{\mu\nu} F^{\mu\nu}] + \ldots$$

(Fun!!) Mathematics, derivatives, integrals, matrices, vectors, functions...

Particle Astrophysics
or
'Looking' at the Stars
Paul Stevenson, Rice University

"We are all
in the gutter
...but some of us
are looking at
the stars"

Oscar Wilde

Particle Astrophysics or 'Looking' at the Stars

Abstract

For millennia humankind has looked at the stars and wondered about the universe. Only in recent years have we been able to "look" at the stars by means other than visible light. Not only can we now use photons of other wavelengths - from radio waves to ultra-high-energy gamma rays - but we can also detect, or try to detect, other particles. These include (i) cosmic rays - charged particles; probably protons, but perhaps heavier nuclei, which arrive from outer space with energies up to 10^{20} eV (the Greissen cutoff, a limit due to collisions with the 3 K microwave background radiation); (ii) neutrinos - which can penetrate through large amounts of intervening matter, whereas photons are blocked by less than 100 grams of matter; (iii) gravitational waves - which can be produced by collapsing binaries of neutron stars or black holes, or other large, violent astrophysical events.

The talk mentioned some of the weird and wonderful detectors now operating or being built: Cherenkov telescopes; air-shower arrays; neutrino detectors, including the AMANDA detector embedded in the antarctic ice; and the laser-interferometer gravitational-wave detector, LIGO. This "New Astronomy" is just beginning. We have already "seen" neutrinos coming from our sun, and the burst of neutrinos that came from the nearby supernova in 1987. We have also "seen" gamma rays of TeV energy - higher than the most energetic photon beam yet produced on earth - coming from the Crab Nebula (the remnant of a supernova that exploded in the year 1054). Many more exciting discoveries can be confidently expected in the next few years.

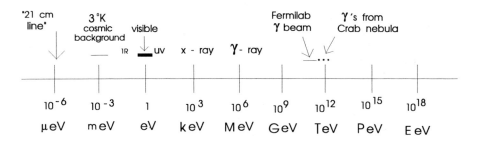

Photon Energy

38 Particle Astrophysics or 'Looking' at the Stars

"Looking" at the stars in other ways

    Ultra-high energy γ's

        Cosmic rays -   charge particles
                                    protons and nuclei

Neutrinos ν

Gravitons

We have seen TeV (10^{12}eV) photons from the Crab Nebula (SN in 1054)

We have seen ν's from the sun.

We have seen ν's from SN 1987A.

Plans to do "neutrino astronomy"

Plans to look for gravitational waves.

Some Weird and Wonderful Detectors

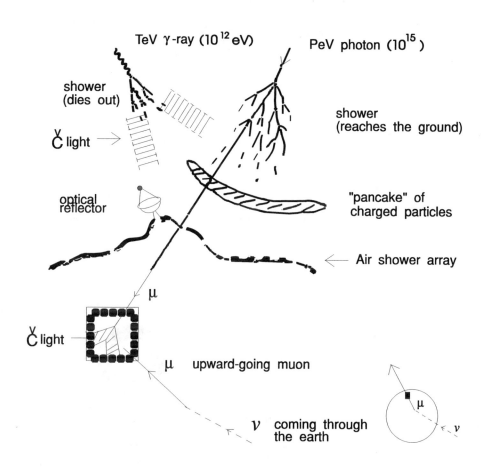

Also:

AMANDA — strings of phototubes deep in the ice at the South Pole

LIGO — Gravitational-wave detector
4km L - shaped laser interferometer

Fundamental Particles

Spin $\frac{1}{2}$
{ Leptons: $\begin{pmatrix} \nu_e \\ e^- \end{pmatrix} \begin{pmatrix} \nu_\mu \\ \mu^- \end{pmatrix} \begin{pmatrix} \nu_\tau \\ \tau^- \end{pmatrix}$ Charge $\begin{matrix} 0 \\ -1 \end{matrix}$

Quarks: $\begin{pmatrix} u \\ d \end{pmatrix} \begin{pmatrix} c \\ s \end{pmatrix} \begin{pmatrix} t \\ b \end{pmatrix}$ $\begin{matrix} \frac{2}{3} \\ -\frac{1}{3} \end{matrix}$ }

Each particle has a antiparticle, eg. $e^+, \bar{u}$

Spin 1
{
Gauge bosons	g(1-8)	γ	$W^\pm, Z^0$	G
	gluons	photon	weak bosons	graviton
force	color	electromagnetic	weak	gravity
}

Spin 0   Higgs particle (?)

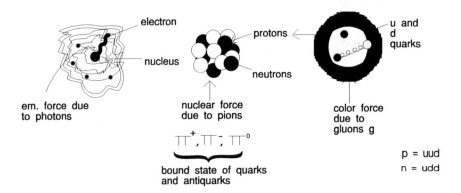

- em. force due to photons — electron, nucleus
- nuclear force due to pions (π^+, π^-, π^0) — protons, neutrons — bound state of quarks and antiquarks
- color force due to gluons g — u and d quarks

$p = uud$
$n = udd$

$\pi^+ = u\bar{d}$
$\pi^- = d\bar{u}$
$\pi^0 = \frac{1}{\sqrt{2}}(u\bar{u} + d\bar{d})$

(Lots of other bound states of quarks "hadrons").

Particle Characteristics

			Mass
P	proton	Interacts strongly Stable	1 GeV
$\pi^{\pm}$	charged pions	Interact strongly Decays in 10^{-8}s. $\pi^+ \to \mu^+ \, \nu_\mu$ $\pi^- \to \mu^- \, \bar{\nu}_\mu$	0·14 GeV
π^0	neutral pion	Interacts strongly Decays in 10^{-16}s. $\pi^0 \to \gamma\gamma$	"
$\mu^{\pm}$	muons	Interact electromagnetically (like a "heavy electron") Decays in 10^{-6}s. $\mu^+ \to e^+ \, \nu_e \, \bar{\nu}_\mu$ $\mu^- \to e^- \, \bar{\nu}_e \, \nu_\mu$	0.11 GeV ($\approx$ 200 times electron mass)
ν	neutrinos	Interact only weakly (but interaction strength increases with ν energy.) Stable (?)	Zero (?) ($\leq 10 eV$)

42   Particle Astrophysics or 'Looking' at the Stars

Cosmic-Ray-induced Air Showers

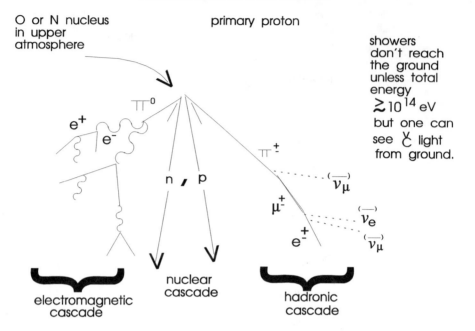

Charged particles with $V > C_{air}$

produce Cerenkov radiation at angle θ

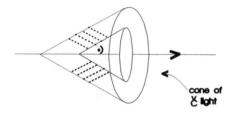

(Almost all particles in the cascade have energy >> their rest mass, so they have v very close to c.

$$E = \frac{1}{\sqrt{1 - v^2/c^2}} \, m_o c^2$$

"Fly's Eye" and Whipple Observatory

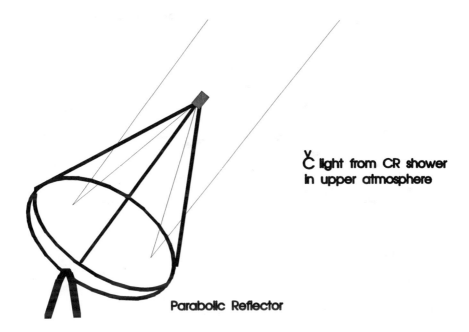

Č light from CR shower in upper atmosphere

Parabolic Reflector

Has seen TeV (10^{12}eV) γ's coming from direction of the Crab Nebula

(Supernova that exploded in 1054, seen by Chinese astromers)

From radio astronomy we know that the Crab contains a pulsar

neutron star - collapsed remnant of original star.

CASA Chicago Air Shower Array

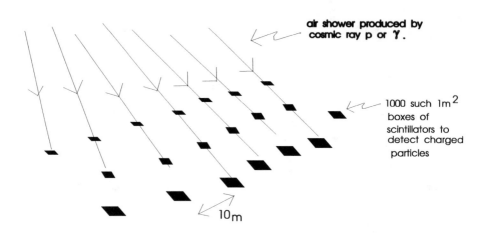

Can measure total energy = E of primary CR.

Precise timing allows <u>direction</u> to be determined.

- to $1°$! (Can see 'shadow' of sun and moon!)

By identifying muons one can distinguish <u>p-induced</u> (contains many μ's) and <u>γ-induced</u> (contains few μ's) showers.

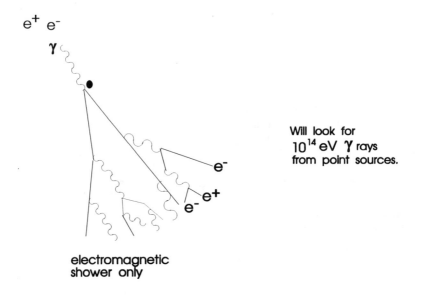

Will look for 10^{14} eV γ rays from point sources.

Flux of Photons

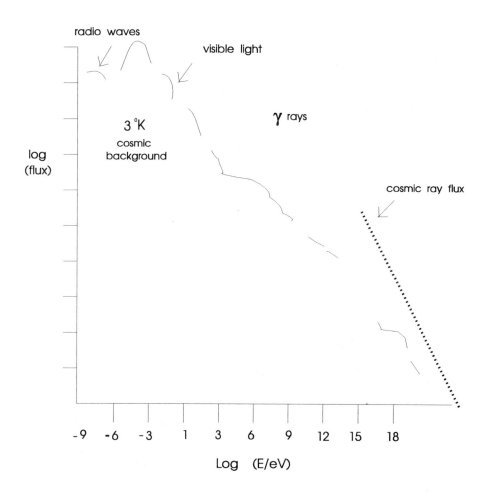

Flux of Cosmic Rays

protons or nuclei

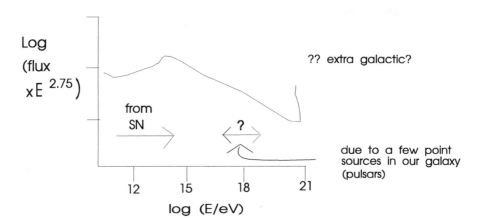

Charged particles are deflected by the

galactic magnetic field $\simeq 10^{-10}$ Tesla

For $E \leq 10^{18}$ eV they will be

confined inside the galaxy

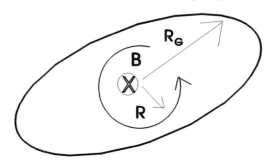

$$R = \frac{E}{eBc}$$

$$R_G \simeq 10^{20} \text{ m}.$$

Direction gets "scrambled."
Direction of arrival at Earth tell us nothing about the direction they came from.

We don't expect to see cosmic ray p's above 10^{20} eV.

"Griessen cut off"

Due to collisions with the 3 K background γ's.

P $\longrightarrow$ •〰〰 $\longleftarrow$ photon from 3 °K background

10^{20} eV $\approx 3\times10^{-4}$ eV

center of mass energy $= \sqrt{m_p^2 + 4E_\gamma E_p}$

becomes $\simeq m_p + 0\cdot1 GeV$

$\approx$ enough energy to create a pion

get $\gamma\, p \to p\, \pi^0$ scattering (or $\pi^+ n$)

Big cross section
  (Δ resonance)

In our reference frame

p has lost ~ 80% of its energy.

48 Particle Astrophysics or 'Looking' at the Stars

Flux of cosmic rays ⇒

in <u>our</u> galaxy, 10^{34} protons are being accelerated to 10^{15} eV every second.

System of size R, involving velocities ~ V, and having a magnetic field B could produce energies up to

$$E \simeq eBRv \qquad (eBv = \text{force}$$
$$\text{force} \times \text{distance} = \text{energy})$$

just from dimensional analysis.

Supernova shock wave mechanism

B ~ 3×10^{-10} Tesla   R ~ 150 light years

v ~ c = 3×10^8 ms^{-1}

⇒   max E $\simeq 10^{17}$ eV

Pulsar mechanism

B ~ 10^8 Tesla   R ~ 10 km

$$v = Rw = \frac{R\, 2\pi}{period} \quad \text{(can be as small as } 10^{-3}\text{s)}$$

⇒ max E $\simeq 10^{19}$ eV

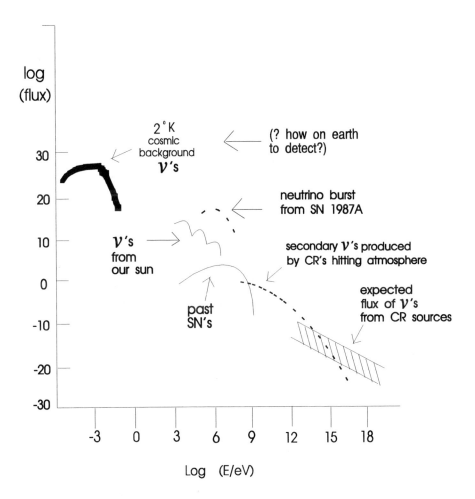

IMB and Kamioka

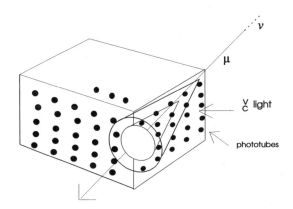

Big tank of water deep underground. (8000 tons)

(Shielded from cosmic rays, except high-energy μ's and ν's).

Saw burst of ν's from SN 1987A

ν's arrived about same time as γ's
$\Rightarrow$ travel at c $\Rightarrow$ massless, or nearly so

ν's from our sun:

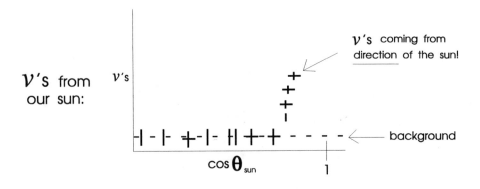

AMANDA Antarctic Muon and Neutrino Detector Array

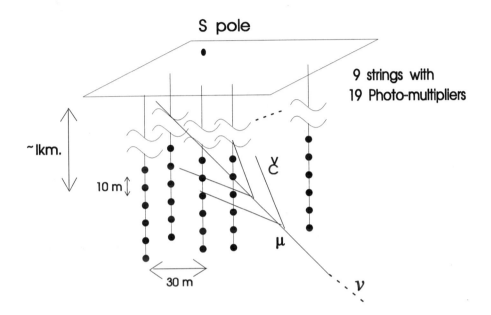

Look for C light from μ

Ice at 1 km depth very transparent
   - free of radioactivity that would give an unwanted background

Aim is to do neutrino astronomy

   ν's come from very high-energy processes.

   Neutral - not bent by magnetic field can penetrate through intervening matter.

   (Photons blocked by $\leq$ 100 grams matter.)

LIGO Laser Interferometer Gravitational-Wave Detector

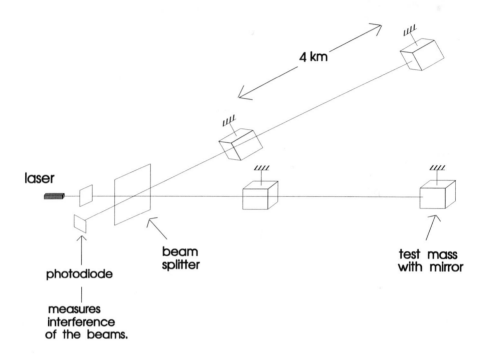

Gravitational wave would distort distances between test masses.

Sensitive to $\Delta L/L \sim 10^{-23}$ (!)

Two or more such detectors to be built.
(one in Livingston, LA and one in Hanford, WA)

Look for coincidences between them (to avoid spurious signals due to random jitter).

Can look for g-waves from collapsing binaries of neutron stars and/or black holes.

Further Reading

Particle Physics and Cosmology
Collins, Martin, and Squires   Wiley-Interscience   0-471-60088-1

The New Astronomy
Halzen, Proceedings of APS Particles and Fields meeting,
    Vancouver 1991

Particle Astrophysics
Sadoulet and Cronin, Physics Today
    April 1991 (Special issue on astrophysics)

Supernovae
    Bethe, Physics Today, September 1990

Cosmic Background Radiation
    Ehrlich, American Journal of Physics, June 1992

LIGO Science, 17th April 1992, Vol., p.325

Dark Matter
    Tremaine, Physics Today, February 1992

The First Three Minutes
    Weinberg, Bantam paperback

TRENDS IN DATA ACQUISITION INSTRUMENTATION

George J. Blanar, Ph.D.
LeCroy Corporation
700 Chestnut Ridge Road
Chestnut Ridge, New York 10977-6499

ABSTRACT: Particle physics research demands unique data acquisition instrumentation in terms of speed, size, cost and architecture. This paper will focus on principal issues related to trends in high-speed, large-scale, economical, sophisticated instrumentation for high energy physics, heavy ion, nuclear and atomic physics as well as large scale astronomical experiments. Examples will be taken from experiments at many national laboratories including BNL, FNAL, CERN, SLAC, etc. as well as LeCroy Corporation's 26 year history in the field of physics research instrumentation.

Finally, instrumentation needs for the next generation of high energy, hadron colliders including the Superconducting Super Collider (SSC) and the Large Hadron Collider (LHC) at CERN will be reviewed and compared to current technologies.

INTRODUCTION

Exponential Problems of Data Acquisition

Physics research electronic instrumentation requirements for data acquisition, triggering and support functions have grown dramatically in the last quarter century. The introduction of higher energy, higher beam intensity facilities has forced the number of channels of data acquisition to expand exponentially. Figure 1 shows this growth with respect to the number of ADC (analog to digital converter) or TDC (time to digital converter) channels for a particular detector of a high energy physics experiment (for example, hadron calorimeter or muon spectrometer), an entire heavy ion experiment, or an entire nuclear physics facility. In all cases, the trends are clear with only a scale change necessary to actually compare disciplines.

Since experimental particle physics has not had an exponentially expanding budget, this growth has been fortunately balanced by an exponential suppression in the price per channel of the data acquisition electronics. Figure 2 plots the cost per channel of many of the most popular commercial LeCroy ADCs and TDCs versus their period of peak use in experimental physics programs.

Until now, the physics community has achieved this cost reduction with the introduction of standards to the data acquisition problem. Standards mean that instrumentation may be configured and constructed based on money saving principles including economies of scale, common crates, power supplies and interfaces, common software libraries etc. Therefore, the same FASTBUS IEEE-960 multi-hit TDC for Drift Chambers can be used by a heavy ion experiment at BNL, or a hadron collider group at FNAL, or an electron collider group at SLAC, or CERN.

The primary reason for the balance in density and cost has been the tailoring of the technology to the fixed parameters represented by both the standard itself and the scope of the instrumentational needs it must cover. Figure 3 reminds us of the exponential trends described above but notes the dates of the introduction of a number of data acquisition standards and the technology they exploited.

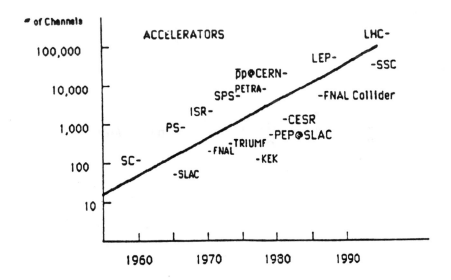

Figure 1
Start-up dates for several Particle Physics Research
Facilities vs. Number of Channels of Data Acquisition Electionics

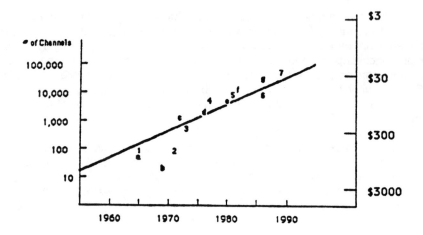

LeCroy Instrument Notation

ADCs	Model#	TDCs	Model#
a	143A	1	108H
b	243	2	226
c	2248	3	2226
d	2249	4	2228
e	4300	5	4303
f	2282	6	1879
g	1882	7	HTD161

Figure 2
Number of Channels/Detector/Experiment and Cost/ADC or TDC Channel vs Time

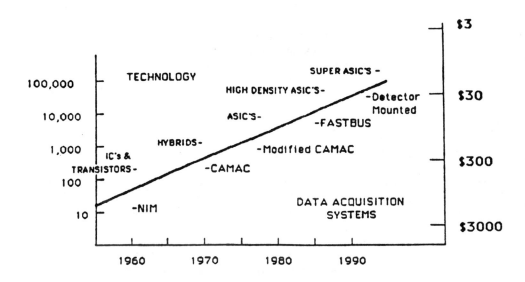

Figure 3
Data Acquisition System Standards and the technology they use

THE EVOLUTION OF STANDARDS

NIM, the First Standard

NIM standard instrumentation which was introduced in the 60's, used discrete components and standard TTL logic integrated circuits to provide the first true international instrumentation standard. It included mechanical standards for crates and modules as well as power supplies. NIM, used primarily for trigger logic, was never considered as a mainstream data acquisition system since it had no standardized computer interface to pipe the acquired data through.

CAMAC, the first C is for Computer

CAMAC (IEEE-583) represented a huge step forward in the 70's. It is defined several architectures for control and data flow. The data flow was optimized to the standard data acquisition computer of the 70's, usually the "mini" such as Digital Equipment Corporation PDPs, Norsk Data NORDs or Data General NOVAs. With its common instrumentation control language ("CNAF"), standardized modules and similar computers, laboratories like CERN and FNAL could develop sophisticated software libraries.

CAMAC also supported the idea of multi-channel data acquisition modules with over 300 channels per crate. This increase in density was made possible through the use of hybridized circuits. Thick film hybrids eventually became the preferred manufacturing technique and permitted not only an enormous savings in the "real estate" that a circuit occupied, but also high performance in terms of noise isolation and speed.

Modified CAMAC, a Standard of Necessity

By the end of the 70's many of the limitations of CAMAC began to affect its use in the higher performance experiments. The most restrictive was the limit on the number of crates that could be run off of one parallel branch (7) and the speed of the transfers (1 μsec/word). Performance was also lost to the 24-bit data format that did not fit well to either the dynamic range of the instrumentation or the word size of the popular acquisition computers (DEC VAX).

In the absence of an IEEE or ESONE standard, many laboratories and commercial companies offered modifications of CAMAC that expanded both the architecture and the data transfer speed. With these changes, Modified CAMAC also benefited from the introduction of Application Specific Integrated Circuits (ASICs) to increase the number of channels of data acquisition that could be contained in a single slot of CAMAC; up to 32 TDCs and even 48 channels of ADCs for example. These single and dual channel ASICs were originally done in semi-custom bipolar and CMOS processes with feature sizes starting at 10 μm and eventually coming down to 4 μm.

FASTBUS, the Physics Community's Own Standard

In the beginning of the '80's, the IEEE-960 FASTBUS standard was established to "tame" the problems stirred by CAMAC. Primarily driven by the labs and the particle physics community in general, problems of architecture, speed, flexibility and in-line computer processing were all accommodated. The cost per channel question was addressed by the unpublished goal of packing 10,000 channels of ADC or TDC in a single rack of electronics. This was roughly the number of channels in the initial design of CERN LEP or FNAL Tevatron detector subsystem. Therefore, a data acquisition system had to have 4 or 5 crates per rack, 20 to 25 data acquisition modules per crate and finally about 100 channels per slot.

Single or dual channel ASICs were insufficient for the FASTBUS requirements. Higher density (4 channels or more per die) ASICs were now required. This push also meant a change from semi-custom processes to full custom designs. This requirement had been felt earlier since the problems of testing these semi-custom, high performance (15-bit dynamic range, 12-bit resolution, 5 psec jitter) devices was difficult using VLSI industry tools. In addition, migrating the design from an engineering to production run often was more difficult than the initial design.

Custom ASICs Were the Key

By the time CDF at the Tevatron and the four LEP experiments started installing their detectors and pulling cables, the FASTBUS standard with 96 channel

ADCs and TDCs had already proven itself in several high rate experiments. These instruments were designed around higher density, 4-channel completely custom ASICs, as shown in Photograph 1.

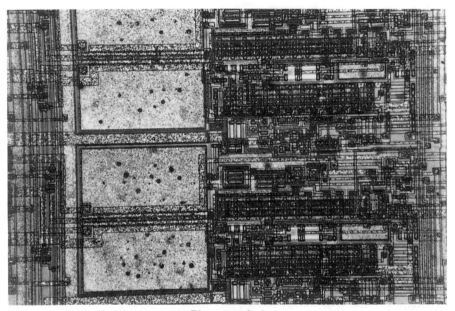

Photograph 1
Micro-photograph of the 4-Input Charge Multiplexer
used in the 15-bit, 96-channel LeCroy FASTBUS ADCs

With time, the various roles of different semiconductor processes were more clearly understood. CMOS worked well for high-speed digital designs like scalers, digital delays, data compression, etc. On the other hand, bipolar lent itself to high performance analog designs including amplifiers, comparators, charge converters, multiplexers, etc. Experiments were tried with exotic processes (for the time) like SoS and GaAs with mixed results. Sometimes it was found that the exotics (SoS for example) were not that well understood by the semiconductor foundries themselves, and their intensive application in physics instrumentation was premature. Others were more expensive than originally foreseen with lower yields (GaAs).

DENSITY IS NOT THE ONLY STORY

Speed and High Rate Trends

Before turning to the next generation of experiments and their requirements, two other performance parameters must be traced for their trends and directions: Triggering and Data Compaction. The former is critical since the integration of the trigger into the data acquisition system is a natural consequence of both the increase in background reactions that the trigger must suppress and the increase in the number of detector channels. Other aspects of the problems include the disappearance of separate detector elements to form the trigger as well as the need to get both the data and the trigger functions out of the tightly packed detector systems.

The Integration of Trigger Systems

Figure 4 illustrates the merging of the trigger systems into deeper levels within the data acquisition system. This trend is driven by the increases in the background rate from the higher energy beams, the higher luminosity of the new accelerators and the corresponding lower relative cross sections of interesting physics phenomena.

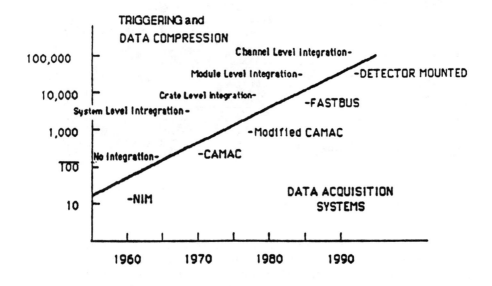

Figure 4
Triggering and Data Compression applied to finer and finer levels of the data acquisition system

NIM systems were usually completely separate from the data acquisition system with a single cable running between them signaling a COMMON START to the TDCs or a GATE to the ADCs. The situation was similar at the start of the CAMAC era but eventually provisions were made to pick off some data at the point where the data acquisition system met the computer (programmable branch drivers) in order to make a third level trigger decision. Deadtime in the computer could then be avoided if the trigger conditions were not met.

With the increases in rates and sizes that were possible with Modified CAMAC, many systems inserted a trigger port at the custom crate controller level to supply data to second level triggers. The LeCroy PCOS III Multi-Wire Proportional Chamber System is probably the most successful example of this implementation and is in widespread use even today.

Finally, FASTBUS provided a backplane wide enough (especially with the auxiliary backplande to allow data from individual modules and even groups of channels within the modules to contribute to second level triggers.

Data Compression, the Key to High Throughput

Figure 4 is also a record of the integration of data compression applied to finer and finer levels of the data acquisition system. This solution was dictated by higher rates and larger amounts of data that could not be simply handled by increasing the width and speed of the data acquisition backplanes. In addition, the speed and bulk of the data had to be reconciled with only a limited amount of computer archive capability.

NIM and early CAMAC systems had relatively few detector channels to read out. Therefore, no compression was necessary. However, with larger CAMAC systems the data acquisition system/computer interface was given the job of compressing data by excluding un-hit TDC channels and ADCs below pedestal.

Once again, Modified CAMAC used the crate controllers to encode the data so that while all the data was read out of the individual modules, only interesting data was passed on the branch. FASTBUS originally used the same approach but recent instruments including the LeCroy Model 1871 Rapid Encoding TDC and others compact data at the module level. Therefore, the backplane bandwidth of the crate is not compromised.

The next generation of data acquisition electronics will have to provide trigger information and compress data on an individual channel basis in order to cope with the demands of both interaction rates and data volumes.

ELECTRONICS FOR FUTURE COLLIDERS

The Case for Detector Mounted Electronics

The next generation of hadron colliders presents us with a critical problem. The late 90's will see the need for several hundred thousand channels per detector device at a cost requirement of less than $10 per channel. Additional restrictions are detailed below, but these two are enough to seriously abandon the idea of a branch/crate/module type CAMAC, VME, FASTBUS or VXI system. The problem is that such standard systems carry an overhead for cooling, power, mechanics, support services, etc. of approximately $10 per channel. With a $10/channel total average budget for the million SSC/LHC channels, this overhead is not acceptable. The most direct solution is to mount the electronics directly on the instrument and integrate the entire electronic chain into the logical and physical structure of the detector.

SSC/LHC, Additional Restrictions and Problems

The list of additional problems and constraints that have to be accommodated is impressive even if compared to the most sophisticated experiments running at LEP or the Tevatron: Beam Crossings every 16 onseconds as opposed to several μseconds, Flight times of a $\beta = 1$ particle through the detector of up to 30 nanoseconds, timing accuracies of multi-hit electronics for drift chanbers to 0.5 nsec, ADCs to measure calorimeter information with 20 bits of dynamic range and 14 bits resolution, Interaction rates a thousand times higher than presently handled, Data per event ten times larger, etc.

The "zeroth" level trigger is always provided by the beam crossing. In order to create time to make a first level trigger, all the possibly useful data must be stored in an analog or digital pipeline for approximately 1 or 2 μseconds. Note that CCDs will not work for this storage element since the clocking currents for them would require the equivalent of an FM radio transmitter in the middle of the detector. Switch capacitor arrays are a more likely solution.

Trigger information must then follow its own routing to be used in a first level trigger decision. This pipelined process determines if the data in the storage pipelines should be routed to the next level of buffering or flushed. If the original storage pipelines are analog, data has to be converted. The second level trigger uses more data itself as opposed the trigger signals. This next level of buffering will probably be organized as a FIFO with the capacity of holding an event for about 5 μsec.

There have been a number of studies on both the intensity and type of radiation background the electronics will have to tolerate. It is not clear that the data used by the defense industry to set their standards of radiation hardness are relevant. However, one conclusion can be drawn: it will be hotter than anything that has been instrumented in elementary particle physics experiments to date.

Estimates range from 10MR for the inner detector to 1 MR by the muon chambers, for 10 year exposures. This eliminates several integrated circuit processes including simple bipolar, MOS and others. CMOS, special bipolar, GaAs and even SoS are better candidates.

Power will be critical. With a million channels of electronics mounted on or near the detector, power will have to be cut by a factor of 20 from the past levels. The use of appropriate technologies (CMOS) and separate, specific valuded power lines will help. Fiber Optic transmission with the LED transmitters on the chambers will not be acceptable. However, an interesting idea was presented at the 1990 IISSC conference by Rykaczewski, et. al. for the L* proposal with off-chamber mounted light sources and an individual shutter integrated in each channel.

All future collider proposals call for hermetic detectors, with as close to 4π coverage as possible for each layer of detector elements. There is little or no room left for the electronics much less the cables or heat/cooling plumbing.

Therefore, data acquisition electronics will have to use new techniques like flexible circuit boards. Trigger, data compression, data signal processing and multiplexing of the data on the detector will help to keep the cabling to a minimum. The use of glass fiber optic links at this point will be highly desirable to get the data out of the detector and into a computer "farm" for further in-line computing and higher level trigger decisions.

Architectures for capturing data and techniques for designing and implementing electronics will have to change to accomodate new levels of reliability and fault tolerance. Studies of techniques used in the space program, avionics and military should help us get fresh ideas.

Because of space restrictions, power, noise sensitivity and reliability, we will have to continue the trend of designing higher channel density ASICs. Eight to 64 channel designs are now discussed and experience with some of the 128 channel Si Strip preamps must be integrated.

<u>A Couple of Models to Consider</u>

Practical solutions to at least some of these severe restrictions can be found in several older experiments. Proton decay and underground astrophysics detectors have been using detector mounted electronics for many years to read out their hundreds of thousands of streamer tube channels. For example, the LeCroy STOS (Streamer Tube Operating System shown in Figure 5) used on the Mt. Blanc experiment featured on-chamber mounted, limited overhead, data acquisition using bipolar based ASICs to keep the channel price and power low. In addition, the trigger is distributed so that all the channels can contribute to a fast global trigger decision with minimum deadtime. However, there was no data compaction used because the trigger rates for these experiments tended to be very low (proton decay is a very rare process!).

Another example comes from the mid '80's. The BNL Multi-Particle Spectrometer needed to instrument approximately 20,000 channels of image chamber detector with a multi-hit TDC but with a density problem similar to some SSC designs. This group used chamber mounted 16-channel, thick film hybrids (LeCroy HTD 161) to amplify, discriminate, time encode and compress the signals and interface to an output bus. The hybrids (Photograph 2) have a footprint of 38 mm by 90 mm and cost about $25 per channel and could be pipelined.

Photograph 2
LeCroy's 16-Input On-Chamber Multi Hit TDC Hybrid
with amplifier, discriminator, time digitizers and control logic

However, this hybrid/ASIC system had no trigger outputs and dissapated at least 250 mW per channel. The ASICs were not particulary radiation hard and the hybrid used several different incompatible processes including CMOS and three different types of bipolar including ECL. Therefore, it would be impossible to make this design directly in one high density multi-process ASIC.

Decade	Channels	Systems	Technology	Trigger & Data Conversion
1960	50 -500	NIM & DMA	Transistors & ICs	Separate
1970	500 -2,000	CAMAC & MOD. CAMAC	Hybrids & ASICS	Controller Level
1980	2,000 - 50,000	FASTBUS	High Density	Module Level
1990	>100,000	Detector mounted	Super ASICS	Channel Level

CONCLUSIONS

Trends from 25 years and Projections for the next 10...

The table below summarizes the critical points developed in this paper. It reviews the changes in density, standards, technology, sensitivity and speed over the last 25 years. We have seen the field go through at least 4 or 5 clearly identifiable stages. For the next generation of hadron colliders, we will have to go through yet another.

The data acquisition instrumentation problems of the 1990's are very difficult. Solving them will be critical to the success of the SSC/LHC programs. We must learn to use multi-process, high density, low power ASICs or be able to compensate for single process ASICs by using advanced techniques like Silicon hybrids. We must study radiation hardness as it applies to a 40 TeV hadron collision environment as opposed to a nuclear holocaust. We must develop architectures that include several levels of pipelines, that are self sparsifying, perform some digital signal processing and allow for the distribution of trigger information.

The last 25 years have taught us a great deal about data acquisition. By studying this discipline we can see the directions we have to go in for the SSC/LHC. Experienced commercial, laboratory and university groups working together are following these directions and meeting these challenges today so that by the end of 1990, the next generation of experiments will be ready to make their contributions to physics research.

ACKNOWLEDGEMENT

The author gratefully acknowledges the assistance in the preparation of this manuscript by Linda Nelson and the careful editing and critique by John Hoftiezer, both of LeCroy Corporation, New York.

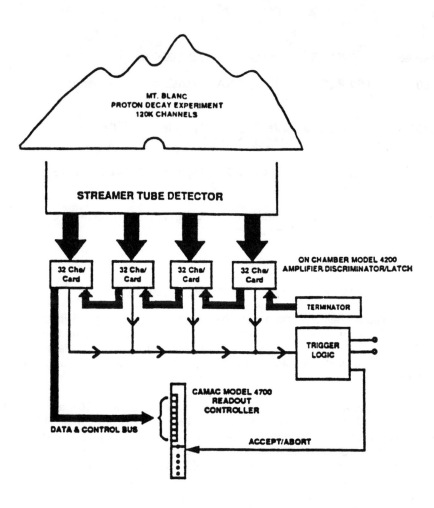

Figure 5
Streamer Tube Operating System Block Diagram

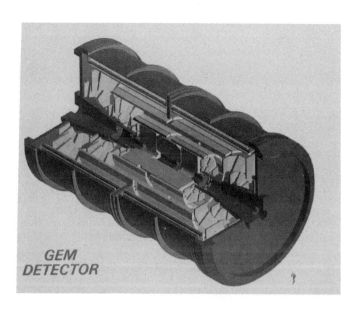

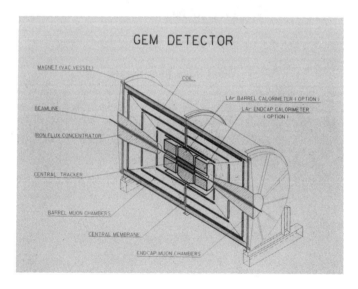

Photograph 3
The GEM detector which will be located at the SSC

Overview of Current High Energy Physics Experiments
Larry Gladney
University of Pennsylvania

<u>Experiments in Particle Physics</u>

* A story composed of questions:

* Where is particle physics done?

* What are the experiments?

* Who does the experiments?

* What do they do?

* Why do they do it?

* How do they do it?

* What have they done?

What are the Experiments?

Who does them?

* Typically, an experiment involves a detector made of several thousand tons of steel and instrumented with thousands of channels of the most sophisticated electronics available.

* Each experiment is managed and used by a collaboration of 50 - 800 Ph.D. physicists, graduate students, and engineers from around the world.

* Experiments take about 10 years to build and are used, in one form or another, for 5 - 20 years.

* Each experiment is capable of doing dozens or sometimes hundreds of different measurements, each fundamental and new by virtue of its environment.

How do they work?

A detector is like the eyepiece on the particle accelerator microscope.
It senses many different particles coming from the particle interaction just as a prism on a microscope senses many different parts of the light spectrum.

Measuring the number and energy of each type of subatomic particle produced allows us to reconstruct the physics at the interaction points.

By simple logic, we know that all unknown physics must manifest itself through the known subatomic particles if it is relevant to the way the world appears to us

The Large Electron Positron Collider

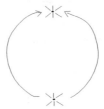

Figure 1

Smashes electrons and anti-matter electrons (positrons) together with 100 billion electron volts of energy at a rate of 50,000 times a second.

There are four experiments at LEP:

* ALEPH
* DELPHI
* OPAL
* L3

Each detector emphasizes detection and measurement of certain types of particles and has, as a central theme, observation/study of the Z^0.

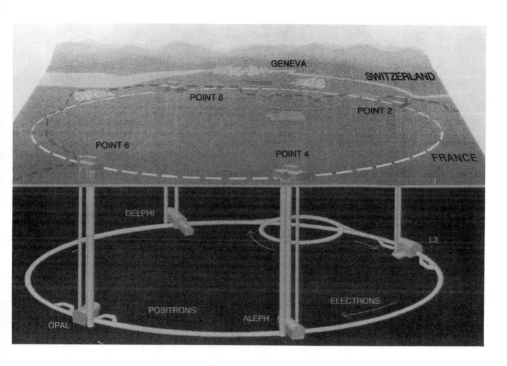

Photograph 1
The Large Electron - Positron Storage Ring

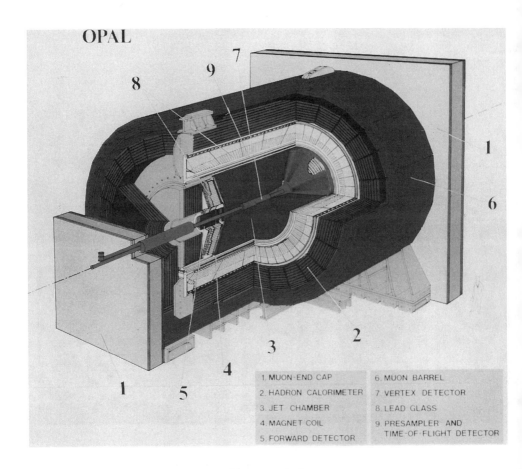

Photograph 2
The OPAL Experiment

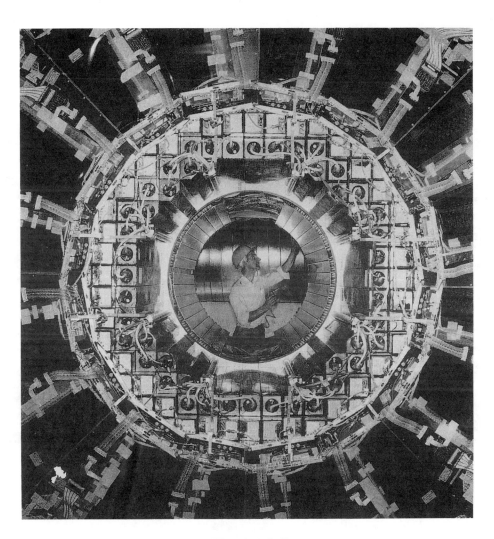

Photograph 3
An end view of the OPAL Experiment

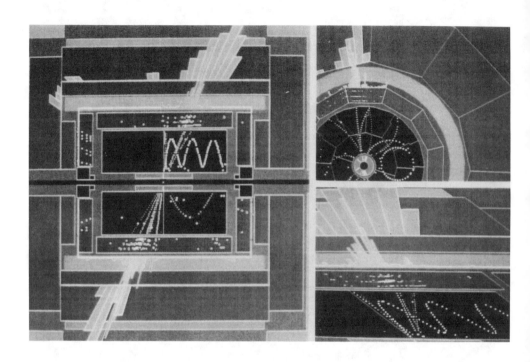

Photograph 4
An $e^+ e^-$ event in the ALEPH detector

The SLAC Linear Collider

Also smashes electrons and anti-matter electrons (positrons) together with 100 billion electron volts of energy, but at a rate of 120 times a second.

Only one experiment can operate at a time.

The Mark II detector took data there in 1990.

SLC is novel because it shows the way to get even greater nergy electron-positron collisions in the future.

Linear Colliders - rifles set barrel-to-barrel

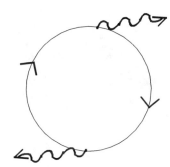

Synchroton radiation
Energy lost

Nothing lost
One pass - one chance for a collision

Figure 2

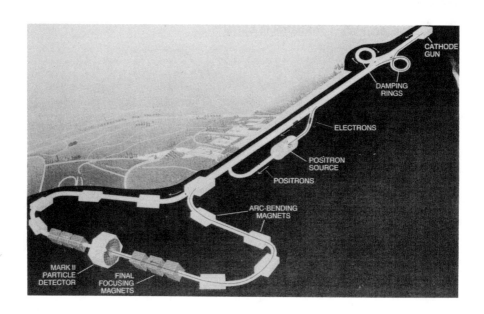

Photograph 5
The Stanford Linear Collider

What has been done?

At present, we believe that all matter can be broken down into fundamental constituents called quarks and leptons. These are grouped into doublets:

$$\begin{pmatrix} up \\ down \end{pmatrix} \begin{pmatrix} charm \\ strange \end{pmatrix} \begin{pmatrix} top(?) \\ bottom \end{pmatrix}$$

$$\begin{bmatrix} e \\ v_e \end{bmatrix} \begin{bmatrix} \mu \\ v_\mu \end{bmatrix} \begin{bmatrix} \tau \\ v_\tau \end{bmatrix}$$

We also believe there are 4 fundamental forces, 2 of which have been mathematically unified:

* Strong force - mediated by gluons; binds quarks to make protons and neutrons, binds protons and neutrons to make nuclei, causes nuclear fission and fusion

* Electromagnetic force - mediated by photons; responsible for atoms, molecules, chemical reactions, light waves, electronics

* Weak force - mediated by W and Z bosons; responsible for neutron decay, beta radioactivity, muon and tau decay, v interactions

* Gravitation - mediated by gravitons (?); determine the large-scale structure of space-'time

What did LEP and SLC do?

One fundamental question that can be asked is:

Are there more families of quarks and leptons?

LEP and SLC say no!

Measuring the number of Z^0s produced as a function of interaction energy tells us the number of families (for light "standard" families) there are.

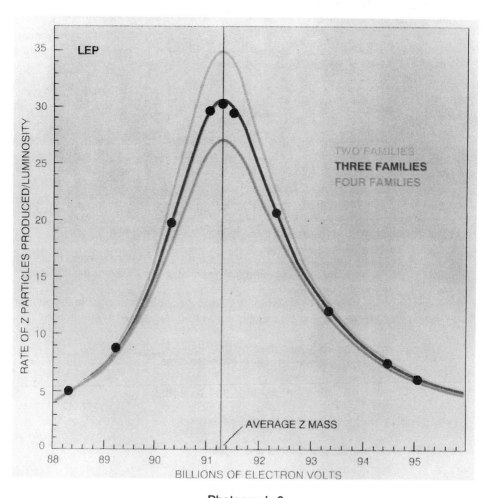

Photograph 6
The width of the Z^0 boson mass compared to the predicted values based on the number of light neutrino families

The Tevatron

A circular, proton-antiproton collider with an energy of 2 trillion electron volts and a collision rate of 50,000 per second.

This is the highest energy accelerator in the world.

There are two colliding beam experiments and many fixed-target experiments.

* Collider Detector at Fermilab (CDF)
* D0 (name of the interaction region)

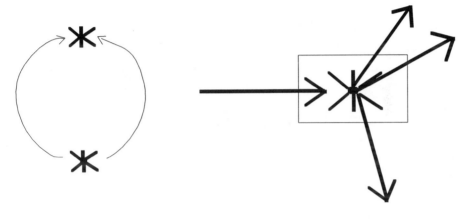

Figure 3

Where is the top quark?

CDF and D0 are dedicated to finding the "last" quark and completing our picture of the Standard Model of EM and weak nuclear interactions.

Without the top, our understanding of nature is quantitatively precise, but qualitatively flawed.

Searching for new particles is the essence of looking for new physics.

Photograph 7
The Tevatron at Fermilab

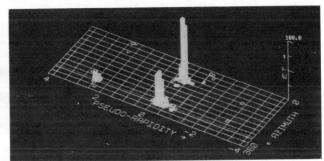

Photograph 8
The CDF Experiment

82 Overview of Current High Energy Physics Experiments

Photograph 9
A $p\bar{p}$ event in the CDF detector

Larry Gladney 83

* Introduction to Protons and Partons

Protons are not as solid as you think!

     Quarks uud

     Glue

     Anti-quarks

Use protons to get quark collisions:

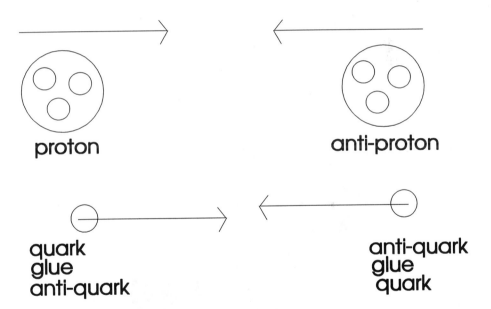

Figure 4

84 Overview of Current High Energy Physics Experiments

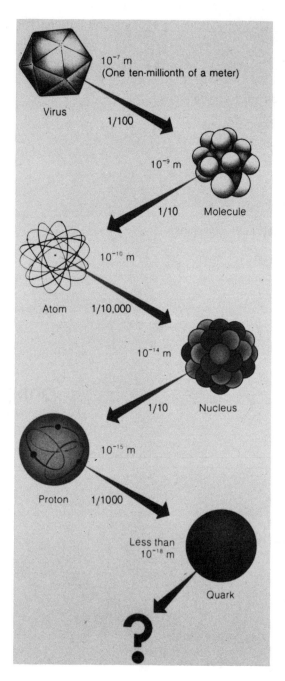

Photograph 10
The structure of matter

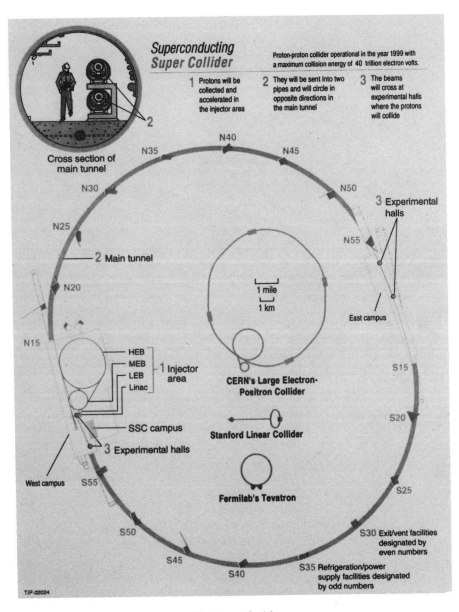

Photograph 11
The **S**uperconducting **S**uper **C**ollider

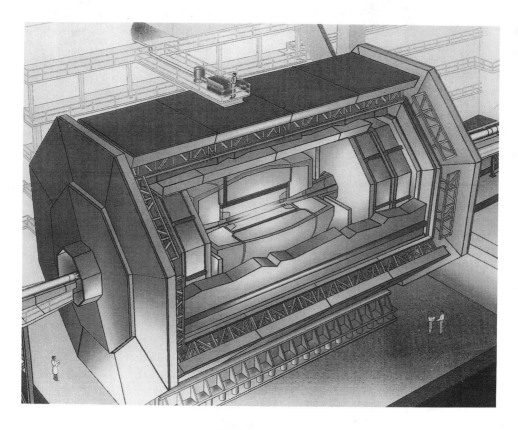

Photograph 12
An artist's rendition of the SDC detector which
will be located at the SSC

1992 Prairie View Summer Science Academy

Participants

Gerardo Acosta	Willowridge High School
Linda H. Ahmed	Bartram Motivation Center
Bret James Akers	The Johns Hopkins University
Nedra Allen	Star Spencer High School
Tynan Anderson	Langston University
Michael S. Anikis	Howard Community College
Julius Barnes	Prairie View A&M University
Joseph M. Beckman	Texas A&I University
Joseph Bezet	Baker High School
Dileep Krishna Bhati	University of Pennsylvania
Althea G. Bluiett	Prairie View A&M University
Billy Bonner	Rice University
Eric Brass	Prairie View A&M University
Mary L. Brass	Jack Yates High School
Lamarr A. Brown	Howard University
Dee Campbell	Prairie View A&M University
Francie Campbell	Prairie View A&M University
Joseph Carlson	The Johns Hopkins University
Robert M. Catchings	Howard University
Tre' Caruthers	Millwood High School
Andrew Chao	University of Pennsylvania

List of Participants

Richard Nelson Cross	Louisiana State University
Kathryn A. Crystle	The Johns Hopkins University
Joe Neil Cunningham	Waller High School
Raffi Curimjee	University of Pennsylvania
Michael T. Davis	Southern University
Della Devine	Prairie View A&M University
William Devine	Prairie View A&M University
Rudolf Douglas	Howard University
Titania E. Dumas	Prairie View A&M University
Alan Wayne Dunlap	Temple High School
Jimmy Emmanuel	Willowridge High School
Danyell Faddis	Jack Yates High School
Ali R. Fazely	Southern University
Joseph Ficht	Willowridge High School
Sheryl Foster	Prairie View A&M University
Byron Freelon	Prairie View A&M University
Wyatt J. Frelot	Star Spencer High School
Sylvester J. Gates	Howard University
Larry Gladney	University of Pennsylvania
Michael Gonzalez	Willowridge High School
Kimberly Guffey	Prairie View A&M University
Burt Joel Guillot	Westfield High School
Deborah Gunter	Langston University
Margaret Haire	Prairie View A&M University
Eric J. Harris	Prairie View A&M University

List of Participants

Franklin Harvin	Walbrook Senior High School
Lee Haynes	Booker T. Washington High School
Patrick Wayne Hervey	High School For Engineering Professions
Lionel Hewett	Texas A&I University
F. Russ Huson	Texas A&M University
Mazharul Huq	Lincoln University
Richard Imlay	Louisiana State University
Mildred Jennings	University of Pennsylvania
Jane Jones	Johnson High School
Rashaan Josey	University of Pennsylvania
Dennis J. Judd	Prairie View A&M University
Barbara Kable	Chopticon High School
George Kalbfleisch	University of Oklahoma
Deborah Keith	D.C. Public Schools
James Klawinsky	Willowridge High School
John Krizmanic	The Johns Hopkins University
Elaine Kuk	University of Pennsylvania
Ronald S. Lee	Westfield High School
James Lindesay	Howard University
Nigel Lockyer	University of Pennsylvania
Ronald L. Loden	Texas A&I University
Raymond David Lofty	Dunbar Senior High School
Reggie Longoria	Texas A&I University
Adetokunbo Lukan	Willowridge High School
Binh Luu	John Bartram High School

List of Participants

Godwin Mayers	Drexel University
Tesfa McGregor	Willowridge High School
Marcia McKeithan	Dunbar Senior High School
Robert Menius	Langham Creek High School
James Metcalf	Louisiana State University
Robert W. Miller	Star Spencer High School
Tonya Montgomery	Southwestern Christian College
Alison Morgan	Lincoln University
Adrian Moore	High School of Engineering Professions
John W. Mosley	Star Spencer High School
Wilfredo Ortiz	University of Pennsylvania
Jason Osborne	Texas A&I University
Kwang Paick	Prairie View A&M University
Rex Peralta	Willowridge High School
Aihud Pevsner	The Johns Hopkins University
Tho Minh Phan	Southern University
Punarvasu Pillalamarri	Camden Catholic High School
Lakshmi Pillalamarri	University City High School
William Pisciella	Booker T. Washington High School
Elmer Ramos	Texas A&I University
Aaron Roane	Howard University
Lynn Roberts	Lincoln University
Sune Rucker	Star Spencer High School
Arletta Saafir	Weatherford Islamic Academy
Kenneth Samuel	Fisk University

List of Participants

Wilson J. Sheppard	Southern University
Inez Simien	Prairie View A&M University
LaBertha Simien	Prairie View A&M University
Elizabeth Anne Smith	Queen Anne's County High School
LaKendrea' Smith	Prairie View A&M University
Doss Sowri	St. Maria Goretti High School
Traci Spiegner	J.O. Johnson High School
Alstene Starks	Star Spencer High School
Ryan R. Stuart	Bartram Motivation Center
Dan Suson	Texas A&I University
Kimani C. Toussaint	University Of Pennsylvania
Christopher Trammell	Southern University
Bao Dinh Truong	University of Pennsylvania
Lon Turnbull	Prairie View A&M University
Billy J. Vegara	Southern University
Jerry Villareal	Texas A&I University
Rafael Villegas	University of Pennsylvania
Shemeka Wade	Booker T. Washington High School
David Wagoner	Prairie View A&M University
Donnell T. Walton	University of Michigan
Carlton Watson	Prairie View A&M University
Berthilla M. Weiss	University of Pennsylvania
Jeremy Whiddon	Baker High school
Chia Yang	Southern University
Te-Hsin (Jimmy) Yang	University of Pennsylvania

AIP Conference Proceedings

		L.C. Number	ISBN
No. 164	Nuclei Far from Stability: Fifth International Conference (Rosseau Lake, ON, 1987)	87-73214	0-88318-364-1
No. 165	Thin Film Processing and Characterization of High-Temperature Superconductors (Anaheim, CA, 1987)	87-73420	0-88318-365-X
No. 166	Photovoltaic Safety (Denver, CO, 1988)	88-42854	0-88318-366-8
No. 167	Deposition and Growth: Limits for Microelectronics (Anaheim, CA, 1987)	88-71432	0-88318-367-6
No. 168	Atomic Processes in Plasmas (Santa Fe, NM, 1987)	88-71273	0-88318-368-4
No. 169	Modern Physics in America: A Michelson-Morley Centennial Symposium (Cleveland, OH, 1987)	88-71348	0-88318-369-2
No. 170	Nuclear Spectroscopy of Astrophysical Sources (Washington, DC, 1987)	88-71625	0-88318-370-6
No. 171	Vacuum Design of Advanced and Compact Synchrotron Light Sources (Upton, NY, 1988)	88-71824	0-88318-371-4
No. 172	Advances in Laser Science—III: Proceedings of the International Laser Science Conference (Atlantic City, NJ, 1987)	88-71879	0-88318-372-2
No. 173	Cooperative Networks in Physics Education (Oaxtepec, Mexico, 1987)	88-72091	0-88318-373-0
No. 174	Radio Wave Scattering in the Interstellar Medium (San Diego, CA, 1988)	88-72092	0-88318-374-9
No. 175	Non-neutral Plasma Physics (Washington, DC, 1988)	88-72275	0-88318-375-7
No. 176	Intersections Between Particle and Nuclear Physics (Third International Conference) (Rockport, ME, 1988)	88-62535	0-88318-376-5
No. 177	Linear Accelerator and Beam Optics Codes (La Jolla, CA, 1988)	88-46074	0-88318-377-3
No. 178	Nuclear Arms Technologies in the 1990s (Washington, DC, 1988)	88-83262	0-88318-378-1
No. 179	The Michelson Era in American Science: 1870–1930 (Cleveland, OH, 1987)	88-83369	0-88318-379-X
No. 180	Frontiers in Science: International Symposium (Urbana, IL, 1987)	88-83526	0-88318-380-3

No. 181	Muon-Catalyzed Fusion (Sanibel Island, FL, 1988)	88-83636	0-88318-381-1
No. 182	High T_c Superconducting Thin Films, Devices, and Applications (Atlanta, GA, 1988)	88-03947	0-88318-382-X
No. 183	Cosmic Abundances of Matter (Minneapolis, MN, 1988)	89-80147	0-88318-383-8
No. 184	Physics of Particle Accelerators (Ithaca, NY, 1988)	89-83575	0-88318-384-6
No. 185	Glueballs, Hybrids, and Exotic Hadrons (Upton, NY, 1988)	89-83513	0-88318-385-4
No. 186	High-Energy Radiation Background in Space (Sanibel Island, FL, 1987)	89-83833	0-88318-386-2
No. 187	High-Energy Spin Physics (Minneapolis, MN, 1988)	89-83948	0-88318-387-0
No. 188	International Symposium on Electron Beam Ion Sources and their Applications (Upton, NY, 1988)	89-84343	0-88318-388-9
No. 189	Relativistic, Quantum Electrodynamic, and Weak Interaction Effects in Atoms (Santa Barbara, CA, 1988)	89-84431	0-88318-389-7
No. 190	Radio-frequency Power in Plasmas (Irvine, CA, 1989)	89-45805	0-88318-397-8
No. 191	Advances in Laser Science—IV (Atlanta, GA, 1988)	89-85595	0-88318-391-9
No. 192	Vacuum Mechatronics (First International Workshop) (Santa Barbara, CA, 1989)	89-45905	0-88318-394-3
No. 193	Advanced Accelerator Concepts (Lake Arrowhead, CA, 1989)	89-45914	0-88318-393-5
No. 194	Quantum Fluids and Solids—1989 (Gainesville, FL, 1989)	89-81079	0-88318-395-1
No. 195	Dense Z-Pinches (Laguna Beach, CA, 1989)	89-46212	0-88318-396-X
No. 196	Heavy Quark Physics (Ithaca, NY, 1989)	89-81583	0-88318-644-6
No. 197	Drops and Bubbles (Monterey, CA, 1988)	89-46360	0-88318-392-7
No. 198	Astrophysics in Antarctica (Newark, DE, 1989)	89-46421	0-88318-398-6
No. 199	Surface Conditioning of Vacuum Systems (Los Angeles, CA, 1989)	89-82542	0-88318-756-6
No. 200	High T_c Superconducting Thin Films: Processing, Characterization, and Applications (Boston, MA, 1989)	90-80006	0-88318-759-0

No. 201	QED Structure Functions (Ann Arbor, MI, 1989)	90-80229	0-88318-671-3
No. 202	NASA Workshop on Physics From a Lunar Base (Stanford, CA, 1989)	90-55073	0-88318-646-2
No. 203	Particle Astrophysics: The NASA Cosmic Ray Program for the 1990s and Beyond (Greenbelt, MD, 1989)	90-55077	0-88318-763-9
No. 204	Aspects of Electron-Molecule Scattering and Photoionization (New Haven, CT, 1989)	90-55175	0-88318-764-7
No. 205	The Physics of Electronic and Atomic Collisions (XVI International Conference) (New York, NY, 1989)	90-53183	0-88318-390-0
No. 206	Atomic Processes in Plasmas (Gaithersburg, MD, 1989)	90-55265	0-88318-769-8
No. 207	Astrophysics from the Moon (Annapolis, MD, 1990)	90-55582	0-88318-770-1
No. 208	Current Topics in Shock Waves (Bethlehem, PA, 1989)	90-55617	0-88318-776-0
No. 209	Computing for High Luminosity and High Intensity Facilities (Santa Fe, NM, 1990)	90-55634	0-88318-786-8
No. 210	Production and Neutralization of Negative Ions and Beams (Brookhaven, NY, 1990)	90-55316	0-88318-786-8
No. 211	High-Energy Astrophysics in the 21st Century (Taos, NM, 1989)	90-55644	0-88318-803-1
No. 212	Accelerator Instrumentation (Brookhaven, NY, 1989)	90-55838	0-88318-645-4
No. 213	Frontiers in Condensed Matter Theory (New York, NY, 1989)	90-6421	0-88318-771-X 0-88318-772-8 (pbk.)
No. 214	Beam Dynamics Issues of High-Luminosity Asymmetric Collider Rings (Berkeley, CA, 1990)	90-55857	0-88318-767-1
No. 215	X-Ray and Inner-Shell Processes (Knoxville, TN, 1990)	90-84700	0-88318-790-6
No. 216	Spectral Line Shapes, Vol. 6 (Austin, TX, 1990)	90-06278	0-88318-791-4
No. 217	Space Nuclear Power Systems (Albuquerque, NM, 1991)	90-56220	0-88318-838-4
No. 218	Positron Beams for Solids and Surfaces (London, Canada, 1990)	90-56407	0-88318-842-2
No. 219	Superconductivity and Its Applications (Buffalo, NY, 1990)	91-55020	0-88318-835-X

No.	Title		
No. 220	High Energy Gamma-Ray Astronomy (Ann Arbor, MI, 1990)	91-70876	0-88318-812-0
No. 221	Particle Production Near Threshold (Nashville, IN, 1990)	91-55134	0-88318-829-5
No. 222	After the First Three Minutes (College Park, MD, 1990)	91-55214	0-88318-828-7
No. 223	Polarized Collider Workshop (University Park, PA, 1990)	91-71303	0-88318-826-0
No. 224	LAMPF Workshop on (π, K) Physics (Los Alamos, NM, 1990)	91-71304	0-88318-825-2
No. 225	Half Collision Resonance Phenomena in Molecules (Caracas, Venezuela, 1990)	91-55210	0-88318-840-6
No. 226	The Living Cell in Four Dimensions (Gif sur Yvette, France, 1990)	91-55209	0-88318-794-9
No. 227	Advanced Processing and Characterization Technologies (Clearwater, FL, 1991)	91-55194	0-88318-910-0
No. 228	Anomalous Nuclear Effects in Deuterium/Solid Systems (Provo, UT, 1990)	91-55245	0-88318-833-3
No. 229	Accelerator Instrumentation (Batavia, IL, 1990)	91-55347	0-88318-832-1
No. 230	Nonlinear Dynamics and Particle Acceleration (Tsukuba, Japan, 1990)	91-55348	0-88318-824-4
No. 231	Boron-Rich Solids (Albuquerque, NM, 1990)	91-53024	0-88318-793-4
No. 232	Gamma-Ray Line Astrophysics (Paris-Saclay, France, 1990)	91-55492	0-88318-875-9
No. 233	Atomic Physics 12 (Ann Arbor, MI, 1990)	91-55595	088318-811-2
No. 234	Amorphous Silicon Materials and Solar Cells (Denver, CO, 1991)	91-55575	088318-831-7
No. 235	Physics and Chemistry of MCT and Novel IR Detector Materials (San Francisco, CA, 1990)	91-55493	0-88318-931-3
No. 236	Vacuum Design of Synchrotron Light Sources (Argonne, IL, 1990)	91-55527	0-88318-873-2
No. 237	Kent M. Terwilliger Memorial Symposium (Ann Arbor, MI, 1989)	91-55576	0-88318-788-4
No. 238	Capture Gamma-Ray Spectroscopy (Pacific Grove, CA, 1990)	91-57923	0-88318-830-9

No. 239	Advances in Biomolecular Simulations (Obernai, France, 1991)	91-58106	0-88318-940-2
No. 240	Joint Soviet-American Workshop on the Physics of Semiconductor Lasers (Leningrad, USSR, 1991)	91-58537	0-88318-936-4
No. 241	Scanned Probe Microscopy (Santa Barbara, CA, 1991)	91-76758	0-88318-816-3
No. 242	Strong, Weak, and Electromagnetic Interactions in Nuclei, Atoms, and Astrophysics: A Workshop in Honor of Stewart D. Bloom's Retirement (Livermore, CA, 1991)	91-76876	0-88318-943-7
No. 243	Intersections Between Particle and Nuclear Physics (Tucson, AZ, 1991)	91-77580	0-88318-950-X
No. 244	Radio Frequency Power in Plasmas (Charleston, SC, 1991)	91-77853	0-88318-937-2
No. 245	Basic Space Science (Bangalore, India, 1991)	91-78379	0-88318-951-8
No. 246	Space Nuclear Power Systems (Albuquerque, NM, 1992)	91-58793	1-56396-027-3 1-56396-026-5 (pbk.)
No. 247	Global Warming: Physics and Facts (Washington, DC, 1991)	91-78423	0-88318-932-1
No. 248	Computer-Aided Statistical Physics (Taipei, Taiwan, 1991)	91-78378	0-88318-942-9
No. 249	The Physics of Particle Accelerators (Upton, NY, 1989, 1990)	92-52843	0-88318-789-2
No. 250	Towards a Unified Picture of Nuclear Dynamics (Nikko, Japan, 1991)	92-70143	0-88318-951-8
No. 251	Superconductivity and its Applications (Buffalo, NY, 1991)	92-52726	1-56396-016-8
No. 252	Accelerator Instrumentation (Newport News, VA, 1991)	92-70356	0-88318-934-8
No. 253	High-Brightness Beams for Advanced Accelerator Applications (College Park, MD, 1991)	92-52705	0-88318-947-X
No. 254	Testing the AGN Paradigm (College Park, MD, 1991)	92-52780	1-56396-009-5
No. 255	Advanced Beam Dynamics Workshop on Effects of Errors in Accelerators, Their Diagnosis and Corrections (Corpus Christi, TX, 1991)	92-52842	1-56396-006-0

No. 256	Slow Dynamics in Condensed Matter (Fukuoka, Japan, 1991)	92-53120	0-88318-938-0
No. 257	Atomic Processes in Plasmas (Portland, ME, 1991)	91-08105	0-88318-939-9
No. 258	Synchrotron Radiation and Dynamic Phenomena (Grenoble, France, 1991)	92-53790	1-56396-008-7
No. 259	Future Directions in Nuclear Physics with 4π Gamma Detection Systems of the New Generation (Strasbourg, France, 1991)	92-53222	0-88318-952-6
No. 260	Computational Quantum Physics (Nashville, TN, 1991)	92-71777	0-88318-933-X
No. 261	Rare and Exclusive B&K Decays and Novel Flavor Factories (Santa Monica, CA, 1991)	92-71873	1-56396-055-9
No. 262	Molecular Electronics—Science and Technology (St. Thomas, Virgin Islands, 1991)	92-72210	1-56396-041-9
No. 263	Stress-Induced Phenomena in Metallization: First International Workshop (Ithaca, NY, 1991)	92-72292	1-56396-082-6
No. 264	Particle Acceleration in Cosmic Plasmas (Newark, DE, 1991)	92-73316	0-88318-948-8
No. 265	Gamma-Ray Bursts (Huntsville, AL, 1991)	92-73456	1-56396-018-4
No. 266	Group Theory in Physics (Cocoyoc, Morelos, Mexico, 1991)	92-73457	1-56396-101-6
No. 267	Electromechanical Coupling of the Solar Atmosphere (Capri, Italy, 1991)	92-82717	1-56396-110-5
No. 268	Photovoltaic Advanced Research & Development Project (Denver, CO, 1992)	92-74159	1-56396-056-7
No. 269	CEBAF 1992 Summer Workshop (Newport News, VA, 1992)	92-75403	1-56396-067-2
No. 270	Time Reversal—The Arthur Rich Memorial Symposium (Ann Arbor, MI, 1991)	92-83852	1-56396-105-9
No. 271	Tenth Symposium Space Nuclear Power and Propulsion (Vols. I–III) (Albuquerque, NM, 1993)	92-75162	1-56396-137-7 (set)

No. 272	Proceedings of the XXVI International Conference on High Energy Physics (Vols. I and II) (Dallas, TX, 1992)	93-70412	1-56396-127-X (set)
No. 273	Superconductivity and Its Applications (Buffalo, NY, 1992)	93-70502	1-56396-189-X
No. 274	VIth International Conference on the Physics of Highly Charged Ions (Manhattan, KS, 1992)	93-70577	1-56396-102-4
No. 275	Atomic Physics 13 (Munich, Germany, 1992)	93-70826	1-56396-057-5
No. 276	Very High Energy Cosmic-Ray Interactions: VIIth International Symposium (Ann Arbor, MI, 1992)	93-71342	1-56396-038-9
No. 277	The World at Risk: Natural Hazards and Climate Change (Cambridge, MA, 1992)	93-71333	1-56396-066-4
No. 278	Back to the Galaxy (College Park, MD, 1992)	93-71543	1-56396-227-6
No. 279	Advanced Accelerator Concepts (Port Jefferson, NY, 1992)	93-71773	1-56396-191-1
No. 280	Compton Gamma-Ray Observatory (St. Louis, MO, 1992)	93-71830	1-56396-104-0
No. 281	Accelerator Instrumentation Fourth Annual Workshop (Berkeley, CA, 1992)	93-072110	1-56396-190-3
No. 282	Quantum 1/f Noise & Other Low Frequency Fluctuations in Electronic Devices (St. Louis, MO, 1992)	93-072366	1-56396-252-7
No. 283	Earth and Space Science Information Systems (Pasadena, CA, 1992)	93-072360	1-56396-094-X
No. 284	US-Japan Workshop on Ion Temperature Gradient-Driven Turbulent Transport (Austin, TX, 1993)	93-72460	1-56396-221-7
No. 285	Noise in Physical Systems and 1/f Fluctuations (St. Louis, MO, 1993)	93-72575	1-56396-270-5